The Determination
of Ionization Constants

The Determination of Ionization Constants

A Laboratory Manual

THIRD EDITION

Adrien Albert
D.Sc. (Lond.), F.R.I.C.
Fellow of the Australian Academy of Science
Emeritus Professor, Department of Chemistry,
Australian National University
Research Professor,
Department of Pharmacological Sciences,
School of Medicine,
State University of New York,
Stony Brook

E.P. Serjeant
M.Sc. (N.S.W.)
Senior Lecturer in Chemistry,
Faculty of Military Studies,
University of New South Wales,
Duntroon, A.C.T

LONDON   NEW YORK
CHAPMAN AND HALL

7148-8832

CHEMISTRY

First published 1962 as
Ionization Constants of Acids and Bases
by Methuen & Co Ltd
Second edition published 1971
by Chapman and Hall Ltd
11 New Fetter Lane, London EC4P 4EE
Third edition 1984
Published in the USA
by Chapman and Hall
733 Third Avenue, New York NY10017

© 1984 Adrien Albert and E.P. Serjeant

Printed in Great Britain at the University Press,
Cambridge

ISBN 0 412 24290 7

British Library Cataloguing in Publication Data

Albert, Adrien
    The determination of ionization constants. —
    3rd ed.
    1. Ionization   2. Acids   3. Bases
    I. Title   II. Serjeant, E. P.
    546'24   QD561

ISBN 0–412–24290–7

Library of Congress Cataloging in Publication Data

Albert, Adrien, 1907—
    The determination of ionization constants.

    Bibliography: p.
    Includes index.
    1. Ionization constants—Measurement—Laboratory
manuals. I. Serjeant, E. P. II. Title.
QD561.A366   1984   541.3'722'028   84–7497
ISBN 0–412–24290–7

QD561
A366
1984
CHEM

Contents

Preface

Preface

This practical manual is devised for organic chemists and biochemists who, in the course of their researches and without previous experience, need to determine an ionization constant. We are gratified that earlier editions were much used for this purpose and that they also proved adequate for the in-service training of technicians and technical officers to provide a Department with a pK service. The features of previous editions that gave this wide appeal have been retained, but the subject matter has been revised, extended, and brought up to date.

We present two new chapters, one of which describes the determination of the stability constants of the complexes which organic ligands form with metal cations. The other describes the use of more recently introduced techniques for the determination of ionization constants, such as Raman and nuclear magnetic resonance spectroscopy, thermometric titrations, and paper electrophoresis.

Chapter 1 gives enhanced help in choosing between alternative methods for determining ionization constants. The two chapters on potentiometric methods have been extensively revised in the light of newer understanding of electrode processes and of the present state of the art in instrumentation.

In the spectrometric methods chapter, the newer IUPAC symbols replace the old. Also the treatment of Hammett's acidity function has been brought up to date. This chapter concludes with help for the preparative organic chemist, namely a rapid spectrometric titration. This yields an approximate pK, knowledge of which will suggest conditions under which the new substance can be extracted and purified in higher yields.

The chapter on zwitterions and other ampholytes has been largely rewritten to give a clearer presentation, and much new information has been incorporated.

The tables of typical ionization constants, in Chapter 9, have been re-compiled from the most reliable of available values, and many additional substances will be found there. This chapter has a new section on substances which modify their ionization by equilibrating with pseudobases (e.g. acridinium and pyrylium salts, also triphenylmethane dyes). The chapter concludes with a new table of the ionization constants of 370 commonly prescribed drugs and other biologically active substances.

Help in the *interpretation* of ionization constants is provided. It is shown how these are related to solubility (Chapter 5), how the degree of ionization at any pH can be calculated (Appendix V), and how ionization constants can aid in deciphering an unknown structure (Chapter 1).

We advise a beginner, before he tackles any unknown substance, to 'enroll' in a *course* by repeating, in the order printed, each worked example in Chapters 2 and 4, both the practical work and the calculations (only easily procurable substances are needed). Much time can be saved in the calculations if the results are set out as shown in these worked examples.

The first examples require only a few, very simple calculations. We have taken care to define the parameters within which these simple methods suffice, and to state clearly when refinements of calculation are needed. We have introduced the more complicated calculations as gradually as possible, and with full explanatory detail. To lessen the tedium of the lengthier calculations, we have devised computer programs. These are presented in a form which clearly shows the pathways of their derivation, to help those who would like to use a desk (electronic) calculator instead of a computer.

Because this is a practical manual, we have touched only lightly on the theory of ionization (Chapter 1 and Appendices I and II). For theoretical study, E.J. King's book *Acid-Base Equilibria* (1965) and R.G. Bates' *Determination of pH, Theory and Practice* (1973) are recommended. Many useful data will also be found in *Electrolyte Solutions* by R.A. Robinson and R.H. Stokes (1959), and *Solution Equilibria* by F.R. Hartley, C. Burgess and R. Alcock (1980), the latter being particularly helpful with the experimental approach to finding stability constants of metal complexes.

We thank Dr D.D. Perrin for helpful discussions.

The Determination
of Ionization Constants

1 Introduction

1.1 What is meant by 'ionization constants'?

In this book, the term 'ionization constant' means any constant which is used to measure the strength of acids and bases. Although sometimes referred to as 'dissociation constant', this term is vague because ionization forms only a corner of the vast field of dissociation phenomena. Thus substrates dissociate from enzymes, micelles dissociate into monomers, and iodine molecules dissociate into iodine atoms. Many other such equilibrium processes are known, but the majority of these dissociations are not ionizations. On the other hand zwitterions are ionized but they are not dissociated.

A minor objection to the term ionization constants is that the constant for a *base* represents an equilibrium in which one ion gives rise to another ion (see equation 1.3, p. 4). This difficulty disappears if we define ionization constants with reference to hydrogen and hydroxyl ions only.

This chapter *outlines* the concepts discussed later in the book. It is supplemented by a general revision of basic theory in Appendices I and II.

1.2 Why do we determine ionization constants?

Ionization constants reveal the proportions of the different ionic species into which a substance is divided at any chosen pH. Appendix V (p. 203) shows how to calculate the percentage of an acid or a base that is ionized at various values of pH and pK_a. It may be deduced from the equations of Appendix V, and also from Fig. 1.1 that a small change in pH can make a large change in the percentage ionized. This is particularly significant if the values of pH and pK lie close together, as they do near the point of half neutralization.

Information about ionization is useful in many ways. For example, different ionic species have different ultraviolet spectra, and significant spectrophotometry can be done only when the pH is so chosen that only *one* ionic species is present. The ionic species of a given substance differ in other physical properties, and in chemical and biological properties as well (Albert, 1979). Ionization constants, by defining the pH range in which a substance is least ionized, indicate the conditions under which it can be isolated in maximal yield (see Chapter 5) and this information has much value in preparative chemistry.

Ionization constants, too, can help to discover the structure of a newly isolated substance. The first question to ask is: Does the postulated structure predict the experimentally determined pK_a values, when calculated theoretically by the method of Perrin, Dempsey and Serjeant (1981)? If so, a considerable verification of the proffered structure has been achieved. Several other ways in which ionization constants reveal structure will now be mentioned. For a fuller treatment of the subject, see Barlin and Perrin's review (1972).

When tautomerism is possible, the structure with the more weakly acidic proton is favoured because it must have the mobile hydrogen more firmly bonded. Ionization constants determinations first showed that 2- and 4-aminopyridine (pK_a 6·9 and 9·2) had the structures of primary amines in equilibrium with very little of the imine tautomers (e.g. (*1·1*)), because 1-methyl-2-imino- and 1-methyl-4-imino-pyridines had pK_as 12·2 and 12·5, respectively (Angyal and Angyal, 1952). Similar reasoning helped to assign structures to the products obtained by methylating the mono-aminopyrimidines, i.e. whether the exo- or the endo-nitrogen atom was alkylated (Brown, Hoerger and Mason, 1955). Again, the methylation of 4-nitroimidazole (*1·2*) gave 1- and 3-methyl derivatives, whose structures were easily assigned from the large and expected difference in pK_a (− 0·53 and 2·13 units, respectively). (Grimison, Ridd and Smith, 1960).

(1.1)   (1.2)

The presence of *covalently* bound water can be detected in a heterocyclic amine by the anonomously high pK_a values that it engenders, by strengthening bases and weakening acids (Albert, 1976; Albert and Armarego, 1965).

To distinguish between a zwitterion and an ordinary ampholyte, in amphoteric substances, there is no better method than to compare the pK_a values as determined in water with those determined in dilute ethanol (see, further, Chapter 8).

Many pairs of geometrical isomers have had their members correctly assigned by comparison of ionization constants (Pascual and Simon, 1964). Axial

carboxy groups are weaker than their equatorial analogues, and many con-
formational problems have been solved by this knowledge (Barlin and Perrin,
1972). Differences in pK_a values form the basis for separating many chemically
similar substances, such as the various penicillins that arise side-by-side from
a fermentation (see also Chapter 5, p. 103).

1.3 Brief summary of the chemistry of ionization

The Brønsted-Lowry theory (Brønsted, 1923) is the most useful and widely
accepted description of the ionization of both acids and bases (see Appendix I).
The underlying concept of this theory is the definition of an acid as any substance
that can ionize to give a solvated hydrogen ion (i.e. a proton stabilized by
interaction with either the solvent or a substance in solution). Conversely a
base is a substance which can accept a hydrogen ion.

In this book we deal almost exclusively with the determination of ionization
constants in aqueous solutions. Most salts are completely ionized in aqueous
solution, but this is not the case with many acids and bases. Very strong acids
and bases are definable as those completely ionized in the pH range 0–14.
Less strong acids and bases are incompletely ionized in parts of this range as
is calculable from their ionization constants by the equations given at the
head of Appendix V. It should also be noted that, in parts of the above range,
the ions of *salts formed from weak acids* (or *bases*) are partly hydrolysed in
aqueous solution, i.e. those ions are in equilibrium with the corresponding
neutral species. This behaviour is also taken care of in the equations of Appendix
V.

Brønsted (1923) was the first to show the advantage of having the ionization
of both acids and bases (i.e. conjugate acids) expressed on the same scale, just
as pH is used for alkalinity as well as for acidity. For acids the ionization process
is

$$HA \rightleftharpoons H^+ + A^- \tag{1.1}$$

and the ionization constant, K_a, is given by

$$K_a = \frac{\{H^+\}\{A^-\}}{\{HA\}} \tag{1.2}$$

where $\{\ \ \}$ represents the *activity* of each ionic species (in mol litre^{-1}).

For bases the ionization is

$$BH^+ \rightleftharpoons H^+ + B \tag{1.3}$$

and

$$K_a = \frac{\{H^+\}\{B\}}{\{BH^+\}}. \tag{1.4}$$

At a given temperature, the constants expressed by equations (1.2) and (1.4) are thermodynamic quantities also known as *thermodynamic ionization constants* which we shall refer to henceforth as K_a^T. These constants are independent of concentration, because all the terms involved are in terms of *activities* (see Appendix II). Another type of constant that we shall use is *the concentration ionization constant*, K_a^C, which is defined for acids as

$$K_a^C = \frac{[H^+][A^-]}{[HA]} \qquad (1.5)$$

in which square brackets denote the *concentration* (as opposed to the activity) of each ionic species. To yield whole numbers (rather than negative powers of ten, which are hard to remember and clumsy to write), equation (1.5) is generally used in the following form, in which pK_a is the negative logarithm of the ionization constant:

$$pK_a = pH + \log[HA]/[A^-]. \qquad (1.5a)$$

For bases, the corresponding expression is

$$K_a = \frac{[H^+][B]}{[BH^+]} \qquad (1.6)$$

or

$$pK_a = pH + \log[BH^+]/[B]. \qquad (1.6a)$$

The main difference between thermodynamic and concentration constants is that the *activities* of the ions have to be taken care of in calculating the former. These activities compensate for the attraction which ions can exert on one another (ion-pair effects) as well as the incomplete hydration of ions in solutions that are too concentrated. The lower the concentration, the less this interaction becomes until, at infinite dilution, the concentration constant becomes numerically equal to the thermodynamic constant. These differences are dealt with in Chapter 3 (p. 47) where help is given in deciding what allowance must be made for activity effects (p. 48). Equation (1.5) can be used for the sake of simplicity provided that: (a) constants are determined in solutions not stronger than 0·01 molar; and (b) only univalent ions are present.

For the present it need only be noted that the activity of a neutral species (molecule) does not differ appreciably from its concentration, at any dilution; and that pH, as commonly determined, is nearer to hydrogen ion activity than to hydrogen ion concentration, although at low ionic strength ($I = \leqslant 0.01$) these terms do not differ greatly* between pH 2 and 10. Hence $\{A^-\}$ is the only unfamiliar quantity in equation (1.2), because $[HA]$ can be substituted for $\{HA\}$, and $\{H^+\}$ is read from the measuring instrument.

---

*See, further, Appendix III.

1.4 The nature of pK_a values

Ionization constants are small and inconvenient figures and hence it has become customary to use their negative logarithms (known as pK_a values) which are convenient both in speech and writing (see equation 1.5a). Thus the pK_a of acetic acid (4·76) corresponds to the ionization constant $1·75 \times 10^{-5}$. Again, the pK_a of ammonia is 9·26, which is more convenient to use than the ionization constant ($5·5 \times 10^{-10}$). pK_b values for bases (see Appendix I), found only in the older literature, (e.g. 4·74 for ammonia at 25°C) can be converted to pK_a values by subtraction from the negative logarithm of the ionic product of water (K_w) at the temperature of determination. The value of pK_w is 14·17 at 20°C, 14·00 at 25°C, and 13·62 at 37°C (see Appendix IV). Thus

$$pK_a + pK_b = 14·00 \quad at \quad 25°C. \tag{1.7}$$

It is evident that pK_a values are very convenient for comparing the strengths of acids (or of bases). The stronger an acid, the lower its pK_a; the stronger a base, the higher its pK_a.

Table 1.1 gives the approximate pK_a values of some common acids and bases. Acids and bases of equivalent strengths have been placed opposite one another. It will prove advantageous to commit this table to memory in order to have a number of reference points for assessing the significance of new pK_a values. Many more values are in Chapter 9. For help in rapid interconversion of K_a and pK_a see Table 11.1 on p. 195.

Table 1.1   Approximate strengths of some common acids and bases

Acids	pK_a	Bases	pK_a
Hydrochloric acid*		Sodium hydroxide*	
	1		13
Oxalic acid	2	Acetamidine	12
	3	Ethylamine	11
Benzoic acid	4		10
Acetic acid	5	Ammonia	9
Carbonic acid	6	Many alkaloids[†]	8
4-Nitrophenol	7		7
	8		6
Hydrocyanic acid, boric acid	9	Aniline, pyridine, quinoline	5
Phenol	10		4
	11		3
	12		2
Sucrose	13	4-Nitroaniline	1

*Completely ionized in the range pH 0–14.
†Also local anaesthetics and antipsychotics

1.5 The shape of a titration curve

When an acid is dissolved in water it ionizes in accordance with equation (1.1). The addition of hydroxyl ions (in the form of potassium hydroxide, for example) disturbs the equilibrium by combining with the hydrogen ions produced by ionization. This reaction greatly alters the ratio of ionized species to neutral species. For example, if exactly one mole equivalent of sodium hydroxide is added, the acid will be quantitatively converted to its anion (A^-), and hence the solution, if evaporated, will give the sodium salt. All the protons available from the acid will undergo the reaction with hydroxyl ions $H^+ + OH^- \rightarrow H_2O$. However, if only 0·1 mole equivalent of sodium hydroxide is added, then 10% of the amount of acid originally dissolved will be converted to the anion, leaving 90% of the original quantity as the neutral species of the acid (HA). Again, if 0·5 mole equivalent is added, the solution will contain equimolar amounts of (HA) and (A^-). Equation (1.5a) reveals that the pK_a value of the acid may be calculated if the pH is measured under conditions of partial neutralization, because the ratio $[HA]/[A^-]$ is known from the degree of neutralization. Thus when 0·1 mole equivalent has been added, the equation becomes $pK_a = pH + \log(90/10)$, or $pK_a = pH + 0·95$; and for 0·5 mole equivalent $pK_a = pH + \log(50/50)$, i.e. at 50% neutralization the pK_a equals the pH. When plotted, as in Fig. 1.1, these results give a *titration curve*, which is

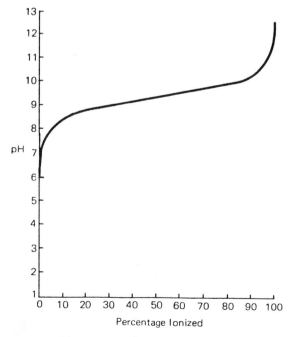

Figure 1.1 A typical titration curve.

approximately sigmoid in shape. Alternatively, if a solution containing a known molar ratio of an acid and its anion is prepared accurately, and the pK_a of the acid is known, the pH of the solution may be calculated.

Similar concepts apply equally to cations which initially establish the equilibrium $BH^+ \rightleftharpoons B + H^+$. For example the anilinium cation (e.g. as its chloride), when titrated with potassium hydroxide in exactly the same manner as acetic acid, yields the same type of pH versus fraction-converted curve (titration curve) shown in Fig. 1.1. The same type of curve will be obtained if a base (e.g. aniline) or an anion (e.g. the acetate ion) is titrated with hydrochloric acid. This titration will alter the ratio of the species involved in an exactly analogous manner, and the relevant mole ratios can be substituted in equation (1.6a) for bases, or equation (1.5a) for anions.

Reference to Tables 2.4 and 2.5, pp. 29 and 30, will confirm that there is no difference in the results obtained (in 0·01 molar solutions) for benzoic acid when titrated with sodium hydroxide or for sodium benzoate when titrated with hydrochloric acid. The latter method, which is essentially the titration of an anion, is advantageous for sparingly soluble acids which may be dissolved in exactly 1 mole equivalent of cold sodium hydroxide; the anion, so formed, can then be titrated with hydrochloric acid. This process, traditionally referred to as a *back-titration*, yields reliable results.

These simple concepts form the basis for the more popular methods available for the determination of pK_a values. In essence these methods involve either:

(a) Measurement of the pH during the stepwise titration of a known weight of substance with accurately standardized hydrochloric acid or potassium hydroxide; this is the *potentiometric method* described in Chapter 2, (see Table 2.3, p. 28 for an example): in this method, the mole ratio of the acid–base conjugated pairs is calculated from the amount of titrant added;

or

(b) Direct observation of the mole ratio of acid–base conjugate pairs in a series of buffered solutions of known pH. A typical use of this concept is the spectrometric method described in Chapter 4.

In both methods, the most reliable results are those obtained from measurements in the range where 20–80% is ionized, because the slope of the titration curve is not too steep (see Fig. 1.1).

1.6 Methods commonly used for determining ionization constants

By far the most convenient method for the determination of ionization constants is *potentiometric titration* described in Chapters 2 and 3. This method requires the measurement of pH using a cell composed of two half cells (each commonly known as an electrode). One of these is reversible to hydrogen ions, that is to say: its potential changes as the hydrogen ion concentration is changed. The other half cell is termed the reference electrode whose potential remains known and invariant throughout; this electrode is used to complete the circuit in the cell. The potential difference between the two electrodes is a measure of the hydrogen ion activity in the solution between the electrodes. The most commonly used electrode system for potentiometric titration is a glass electrode in combination with a saturated calomel electrode used as a reference electrode. A hydrogen electrode, in place of a glass electrode, is not recommended for general use; it is more troublesome, and has often given wrong values due to hydrogenation of the substance being determined. The potential of the hydrogen electrode is nevertheless the thermodynamic standard against which all other potentials are measured and hence its role in the theory of pH is that of a primary standard.

In the pH range 12–14 all types of glass electrode are inaccurate. As a reference electrode, some workers prefer a silver–silver chloride electrode to a calomel electrode. Of the commercially available instruments for measuring pH all are quite suitably equipped with a calomel electrode and a glass electrode. A brief outline of the theory of pH and its derivation is given in Appendix VI.

Should potentiometry prove to be inapplicable, the method of choice becomes ultraviolet *spectrophotometry*, where both the ion and the molecule can be isolated in solution and observed independently. The pK_a is derived from measurements of the *proportions* in which these two species occur, over a range of pH values. This method is described in Chapter 4.

Whereas an ionization constant can usually be determined by potentiometry in 20 min, the spectrometric method may take as much as half a working day unless it is being used routinely. It is particularly suitable for substances too sparingly soluble to be measured by potentiometry, and for those whose values lie outside the recommended range (pK_a 2·0 to 11·0) for potentiometry. The principal requirements are that the substance should absorb ultraviolet or visible light, and that the absorption of the ion should differ from that of the molecule in wavelength or in intensity (and, preferably, in both).

The spectrophotometric method is related to potentiometry in that the spectra are measured in buffers whose pH values have been potentiometrically determined.

It is insufficiently well known that the spectrophotometric method can be adapted for measuring optically *transparent* substances. This is done by adding pH-sensitive indicators in such small amounts that no appreciable effect is exerted on the equilibrium under investigation (see p. 94). Although a more time-consuming method, it is capable of giving reliable results (for examples, see Kolthoff, 1925).

A large variety of methods alternative to potentiometry and spectrophotometry is available, each one having only limited usefulness. The traditional *conductimetry* takes longer than potentiometry, and requires more careful temperature control, but can give very accurate results. Essentially, it measures the changes in conductivity which result when an aqueous solution of the unknown is successively diluted with water. No titration with acid or base is required, but conductimetric titrations have proved useful for very weak acids having a pK_a greater than 11. In general, conductimetry is confined to the measurement of moderately weak acids (pK_a 2–5). It is described in Chapter 6.

Ionization constants can also be measured, although not very accurately, by observing the increase in aqueous *solubility* that accompanies the process of ionization. This and some other phase-change methods (such as *partition* effects) are discussed in Chapter 5.

Raman and *magnetic resonance spectroscopy* methods are described in Chapter 7. They have been used mainly for pK_a values that lie outside the range 2–11. *Thermometric titrations* (Christensen *et al.*, 1966) are noteworthy in that they have furnished pK_a values for the carbohydrates, weak acid of pK_a about 12. Special apparatus and calculations are required (see, further, Chapter 7).

Because it often produced highly erroneous results, no more than historical interest is attached to determining the ionization constant of an acid by measuring its *catalysis of hydrolysis*, e.g. of an ester or glucoside.

1.7 What degree of precision is required?

The degree of accuracy required has been discussed in the Preface. We can now discuss how much precision (i.e. agreement) is necessary to achieve it.

When a set of results has been obtained, the investigator must decide quickly whether they are good enough for his purpose, or whether the determination should be repeated with some change in conditions. In our experience, a set of results (from a single determination) that has a wide scatter, has usually turned out to be even more inaccurate than even the extreme limits suggest. Thus a pK_a of 7.83 ± 0.2 (where ± 0.2 is the scatter or spread) is as likely to lie outside the range 7·63–8·03 as in it. Thus this degree of precision (± 0.2) is not sufficient for accuracy. At the other extreme, an investigator attempting to obtain an ionization constant with less scatter than ± 0.01 may have to spend many weeks on determinations, and this would only be justified if he were trying to establish an international standard for some calibration purpose.

Obviously a middle way must exist. The majority of readers of this book will want to obtain trustworthy values, ones that can be repeated throughout the scientific world. At the same time they will usually want to obtain these values with reasonable speed. From experience we think that *no more scatter than 0·06 in a pK_a value should be allowed* for a set of readings in any one estimation. We shall leave the definition of a 'set' until each technique is discussed.

Scatter is calculated by taking antilogarithms of each pK_a value in a set, averaging these, and writing down the logarithm of the average as the pK_a. The largest deviation between this value and any value in the set is then written after the pK_a as its scatter. Thus if the average is found to be 3·93, and the nine values of the set (of which it is an average) were respectively 3·91, 3·91, 3·93, 3·95, 3·94, 3·96, 3·97, 3·94, 3·91, then the value 3·97 has the largest deviation, namely 0·04, and so the scatter is recorded as $\pm 0\cdot04$ (or $+0\cdot04$ and $-0\cdot02$), and the complete pK_a is reported as $3\cdot93 \pm 0\cdot04$.

The following illustrates the error of averaging logarithms instead of first converting the values to antilogarithms. Addition of two pK_a values, say 2·10 and 2·90, and division of the sum by two gives 2·50 which is wrong, because it is not a mean but a square root! Instead, the pK_a values should be converted to their antilogarithms (126 + 794): these are added and the sum is divided by two, giving 460. The logarithm of 460 is 2·66, and this is the correct average.

Some authors like to report their constants with a 'probable error' instead of a scatter. The use of probable error suggests a smaller deviation than is really the case, and withholds useful information from the reader who wants to know the maximal deviation. The probable error is obtained by subtracting the average pK_a from each value in the set without attention to sign. The sum (Σ) of these deviations is then multiplied by 3 and divided by the product of the number of values in a set and the square root of this number. Thus, in the example given above, 3·93 is subtracted from each of the nine values: the differences added without regard to sign give $\Sigma = 0\cdot17$. The probable error is then:

$$\frac{3 \times 0\cdot17}{9\sqrt{9}} \quad \text{or} \quad \frac{0\cdot51}{27} \quad \text{or} \quad 0\cdot02.$$

The 'order of accuracy', which is the probable error divided by the average value, provides an even smaller number.

In another treatment, known as the 'standard deviation', each difference between a reading and the arithmetic mean of all readings is squared. The nine products are then added and divided by the number of readings. The square root of this quotient is the standard deviation (in the present case $\pm 0\cdot02$).

Of all these alternatives, the scatter gives the reader the most valuable information because it points to, rather than cancels, *outlying* values. It is, however, of no use unless the number of readings in the set is also reported.

1.8 The effect of temperature on ionization constants

Ionization constants vary with temperature; the correlation curve is usually a parabola with a rather flat maximum. For many acids, including all carboxylic acids, this maximum is near 20–25°C, and hence the ionization constants of such substances can be determined without refined temperature control. Thus the pK of formic acid does not increase by more than 0·01 unit between 0° and 100°C. But many non-carboxylic acids are temperature sensitive, for example phenol becomes stronger by 0·012 unit of pK_a for each degree rise in temperature. Some inorganic acids, like phosphoric acid, are temperature-insensitive, whereas boric acid becomes stronger by about 0·006 unit per degree.

Nitrogenous bases, highly temperature sensitive, become weaker as the temperature is increased. Thus the pK_a of aniline is 4·66 at 20°C, but only

4·52 at 30°C. The temperature effect is much greater with stronger than with weaker bases. A general equation, governing the temperature variation for the ionization of monoacidic bases, $BH^+ \rightleftharpoons B + H^+$ (Perrin, 1964) is well supported by experimental evidence. This equation assumes that the standard entropy change for nitrogenous bases is about $-4\,cal\,°C^{-1}\,mol^{-1}$, and this yields the expression for the temperature coefficient

$$\frac{-d(pK_a)}{dT} = \frac{pK_a - 0·9}{T} \tag{1.8}$$

where T is in degrees absolute (K). Values of the temperature coefficients for a number of mono-acidic bases are given in Table 11.7 (p. 200, Appendix IV) where a more detailed discussion is to be found. The use of equation (1.8) is illustrated by the following:

The pK_a of a monoacidic base has been found to be 7·12 at 25°C (298 K) and it is desired to find its temperature coefficient in order to calculate its pK_a at 20°C. Application of equation (1.8) yields the temperature coefficient

$$\frac{-d(pK_a)}{dT} = \frac{7·12 - 0·9}{298} = 0·021.$$

This value is the rate of *decrease* of pK_a as the temperature is *raised* by 1°C. A decrease of temperature from 25°C to 20°C will therefore increase the pK_a value by $5 \times 0·021$ units. Thus the pK_a at 20°C will be 7·22 (i.e. the base is stronger at the lower temperature). Bases which have *two* ionizable groups (diacidic bases) require a different equation to describe the temperature dependence of the pK_a for the process

$$H_2B^{2+} \rightleftharpoons BH^+ + H^+.$$

This represents the ionization process of the *dication*, the weaker of the two groups. This equation is

$$\frac{-d(pK_a)}{dT} = \frac{pK_a}{T}. \tag{1.9}$$

For amino acids the temperature dependence of the basic pK_a is as given by equation (1.8).

From the above it is evident that good temperature control is required in the standardization of a potentiometer with borax, and other non-carboxylic, buffers, and also in the determination of the ionization constants of all bases and many acids.

1.9 Molality and molarity

Investigations of the effect of temperature variation upon an ionization constant are best carried out using *molal concentrations* to avoid the effect of temperature upon concentration. Molality is the number of moles dissolved in *1 kilogram* of solvent rather than in *1 litre* as in the molar scale of concentrations. The use of a top pan balance makes the preparation of molal solutions almost as convenient as preparing their molar analogues. Ionization constants, when determined by using molality, have the dimensions mol kg^{-1} but when determined by using molarity, they have the dimensions mol l^{-1}. For most practical purposes, the difference is inconsiderable; yet molality has the advantage of being independent of temperature.

2 Determination of ionization constants by potentiometric titration using a glass electrode

Potentiometry, because it is most economical of time, is usually the best choice for determination of ionization constants. From Sørensen's announcement, in 1909, of his concept and definition of pH, the means for determining this quantity have steadily improved. All the work described in this chapter can be done with the now universally available potentiometer–electrode assembly, commonly referred to as an Ion Activity Meter or pH set. If used with care, this apparatus can be made to give acceptable results. An outline of the theory of pH is given in Appendix VI (p. 203).

The *hydrogen electrode* is the ultimate standard to which all determinations of pH are referred. The solution to be measured is saturated with hydrogen, and is confined in an atmosphere of this gas. The electrode consists of finely divided platinum (adhering to a platinum plate) which reversibly converts hydrogen gas into hydrogen ions, thus:

$$H_2 \rightleftharpoons 2H^+ + 2e. \tag{2.1}$$

The metal plate takes up the free charge (indicated above by the two electrons, $2e$) from the solution until the electric potential between the plate and the solution has built up so as to prevent further change. This potential is a measure of the tendency of the gas to split into ions and so pass into the solution. If the hydrogen gas is at a pressure of one atmosphere and the potential of the electrode is E (in V), then:

$$E = -\frac{RT}{F} \ln\{H^+\}, \tag{2.2}$$

where T is the absolute temperature, R is the gas constant and F the Faraday. To measure this potential, the platinum plate is connected to the measuring instrument, which is essentially a potentiometer plus a high resistance galvanometer, and the circuit is completed through a second electrode of constant potential, usually a calomel half cell. The pH of the unknown solution is calculated from equation (2.3):

$$\log\{H^+\} = pH = \frac{E_{cell} - E_{cal}}{0\cdot0592} \quad \text{at} \quad 25°C, \tag{2.3}$$

where E_{cell} is the observed potential and E_{cal} is the potential of the calomel electrode under the experimental conditions.

2.1 Apparatus for general use

The hydrogen electrode is, nowadays, seldom used for the determination of ionization constants because it is inconvenient and easily 'poisoned' (irreversibly inactivated). Moreover, it often chemically alters (by hydrogenation) the substance being measured. The more convenient glass electrode has taken its place. This electrode consists of a thin-walled bulb of soft glass whose unusual composition has been specially designed for this use. This electrode is filled with a phosphate buffer of approximately pH 7, containing chloride ion, into which dips a small silver–silver chloride electrode. The circuit is completed with a calomel or silver chloride half cell. Essentially, the potential across a glass membrane varies linearly with the pH of the solution. However, in the high pH region (> 10), the potential developed by the glass membrane becomes sensitive also to alkali metal cations, particularly sodium ion, causing the observed pH readings to be lower than what they should be. Tetramethylammonium hydroxide gives less of this type of error, but is too unstable in solution for convenient use as a titrant. When determining pK_a values of strong organic bases, the errors that arise from inorganic cations can be avoided by simply measuring pH values after addition of an equivalent of hydrochloric acid, in ten equal portions to the free base (Searles *et al.*, 1956).

The potential generated by the hydrogen ions in the solution is measured by an electronic potentiometer assembly (pH, or Ion Activity, Meter). The relationship between the potential of the glass electrode and the pH of the solution has the general form of equation (2.3), but also involves terms which can change daily through asymmetry potentials and other variable effects. Hence this electrode cannot be used as a primary standard, but it does provide a very convenient way of comparing the pH of a series of solutions. To serve this purpose, it is calibrated before and after use with a pair of known buffers, the pH of one of which must lie near to the pH region to be measured.

In general, most ion activity meters manufactured after 1973 are suitable for the measurement of pH during a titration. When used with correctly functioning glass and calomel electrodes employed in solutions maintained at constant temperature, their stability is such that the measured pH values for buffer solutions do not vary by more than 0·02 unit without restandardization over a working day; quite frequently the variation is less than 0·02 unit. This performance far exceeds the requirement of *this* chapter, that the pH recorded for standardizing buffer solutions should not change by an amount greater than 0·04 unit over the time taken for a titration (20–60 min), and actually many pH meters manufactured before 1973 were capable of meeting this specification. At the present time, such a variety of suitable commercial makes are available that it would be invidious to distinguish between them. Solely for the guidance of beginners, we state that the Pye-Unicam 290 Mark 2 pH Meter has been recommended by Spillane *et al.* (1982), and that since 1974 we have used a Philips Ion Meter Type DW 9413. The meter should be mounted on a sturdy bench out of direct sunlight. It will be found convenient, though not essential, to connect the output of the meter to a pen recorder. Confirmation that an equilibrium value of pH has been attained is facilitated greatly by such an arrangement, particularly after the electrodes have been transferred from one solution to another. The tracings also give a day-to-day record of the rate of response of the glass electrode from which it is easy to detect a deterioration in performance.

Automatic titration devices which lack differentiating circuits are to be avoided because they remove human control just where it is most needed to ensure accurate work, Instruments that plot pH against titration volumes, on graph paper, usually employ so small a scale that the available sensitivity of the set is seriously reduced. Such an attachment allows one to see what is going on during the titration, but, for the actual calculation of pK_a, it is essential to rely on the dial readings.

The cell most commonly used for the laboratory measurements of pH consists of a glass electrode, reversible to hydrogen ion, which is used in conjunction with a calomel reference half cell. Both of these should be fitted with leads about one metre long. Nowadays the solution in the reference half cell is either 3·5 or 3·8M potassium chloride which has largely replaced the saturated solution of the older type. The latter solution is no longer considered suitable because lowering the temperature (in storage, for example) usually caused the liquid junction to become clogged with crystals. However, the potentials of all types of calomel reference cells are susceptible to a temperature hysteresis which has been attributed to the disproportionation reaction:

$$Hg_2Cl_2 \rightarrow Hg + HgCl_2.$$

As a result of this, mercury (II) complexes are formed in the potassium chloride reference solution which cause a slow drift in potential. When the temperature is lowered, the solution still contains the dissolved mercury(II) species and thus retardation of equilibrium is accentuated. The amplitude of the effect depends upon the chloride ion concentration of the reference solution, being very small in 0·1M potassium chloride solution but most pronounced in saturated potassium chloride. At high temperatures, the calomel electrode has only a limited life so that it is inadvisable to use it above 70°C for long periods. The type of calomel reference half cell usually supplied for use in measuring pH is often fitted with a porous ceramic plug through which electrical contact is made between the potassium chloride solution of the half cell and the test solution. This plug allows a very slow leakage of the bridge solution (typically 0·01 to 0·1 ml per 5 cm head per day) and this gives rise to a liquid junction potential (see Appendix VI). This design has a reproducibility of $\pm 0·1$ mV ($\pm 0·002$ pH at 25°C) in aqueous buffer solutions and is very suitable for pH titrations in which there is usually no chance that the titrant contains particulate or colloidal material. Such material causes this type of junction to become fouled which makes it very difficult to clean, and causes unstable potentials. It is good practice, therefore, to use a reference half cell of known history and to reserve it exclusively for pH titrations.

In our experience, the response of glass electrodes now available commercially is generally as good as that of a hydrogen electrode in the pH range 3–10. However, a new glass electrode must be conditioned before use by storing it in water for a time long enough to ensure that the gel layer of the electrode is fully formed. We use a conditioning period of 14 days in water. We find that, after this treatment, the potential of the electrode, when immersed in a buffer solution at a constant temperature, is reproducible over a period of days. When the precision aimed at in this chapter suffices, however, satisfactory results may be obtained with an electrode conditioned for only 24 hours, but an unidirectional drift in pH over the period of one hour is likely to be noticed.

Before beginning a titration, the electrodes must be standardized according to the instructions supplied by the manufacturer of the pH meter. The first reference solution to be selected from those listed in Table 2.1 should have a pH value approximately the same as that of the solution contained in the bulb of the glass electrode. Unless stated otherwise, this can generally be assumed to be about pH 7, for which the phosphate buffer of pH 6·86 (at 25°C), given in Table 2.1, is suitable. Each of the electrodes is washed three times with small portions of this buffer solution, delivered from a Pasteur pipette, before dipping them in the buffer solution. After a stable pH value has been reached, the instrument reading is adjusted, by turning the asymmetry potential control, so that a value of 6·86 appears on the dial. For routine work, the electrodes are then immersed in the phthalate reference solution of Table 2.1 maintained at the same temperature. Once the reading has become stable, the slope control is adjusted until 4.00 appears on the dial. The electrodes are then returned to the phosphate buffer solution and the whole procedure is repeated until no further adjustments are required. The electrodes are then transferred from the phthalate to the borax reference solution and the equilibrium pH value of this solution measured. The value obtained should not differ from the published value by more than 0.02 pH unit. Provided this condition is met, the electrodes are suitable for determining pK_a values. However, if the value obtained for the

Table 2.1   pH Reference Solutions of the U.S. National Bureau of Standards[t][a]

Composition and molality	pH at 25°C
Primary reference solutions:	
A. Potassium hydrogen tartrate, saturated at 25°C	3·557
B. Potassium dihydrogen citrate, 0·05m	3·776
C. Potassium hydrogen phthalate, 0·05m	4·004
D. Potassium dihydrogen phosphate and disodium hydrogen phosphate, each 0·025m	6·863
E. Potassium dihydrogen phosphate, 0·008 695m, and disodium hydrogen phosphate, 0·030 43m	7·415
F. Sodium tetraborate (borax), 0·01m	9·183
G. Sodium bicarbonate, 0·025m, and sodium carbonate, 0·025m	10·014
Secondary reference solutions	
H. Potassium tetroxalate, 0·05m	1·679
I. Tris, 0·016 67m, and tris[b] hydrochloride, 0·05m	7·699
J. Calcium hydroxide (saturated at 25°C)	12·454
From other sources	
K. Sulphamic acid (1·0% w/w)[c]	1·18
L. Disodium hydrogen phosphate (1·419 g) and 0·1N sodium hydroxide (100 ml) in water to 1 litre[d]	11·72

[a]International Union of Pure and Applied Chemistry (1978).
[b]Trishydroxymethylaminomethane.
[c]Spillane *et al.* (1982).
[d]Bates *et al.* (1950).

borax solution is lower than the published value (9·18) by more than 0·02 pH unit, then it is probable that the glass electrode is worn, and needs to be replaced. This is particularly likely if the instrument came to equilibrium only slowly in this borax test.

It is difficult to define a slow rate of response for the glass electrode, in the range 20–25°C, because the normal rate depends on the type, thickness, and quantity of the glass used in forming the membrane. Electrodes of high resistance, and miniaturized electrodes, have a slower response than average types but, in general, the time taken to reach an equilibrium value of pH on transferring the electrodes from pH 4·0 to pH 9·2 buffer solutions should not exceed 8 min, provided the solutions are at the same temperature. The recorded pH is usually stabilized to within 0·002 pH unit of the eventual figure during this 8 min period. However, at temperatures of 15°C and lower, the resistance of the membrane increases and the response becomes slower.

Although the method of standardization given above suffices for the accuracy and precision aimed at in this chapter, greater accuracy can be attained by selecting two reference solutions the pH values of which lie on either side of the pH values expected to form the central portion of the titration. These values may

Table 2.2   Effect of temperature on pH of solutions selected* from Table 2.1

Solution	Temperature (°C)					
	20	25	30	35	40	50
A	—	3·557	3·552	3·549	3·547	3·549
C	4·002	4·008	4·015	4·024	4·035	4·060
D	6·881	6·865	6·853	6·844	6·838	6·833
E	7·429	7·413	7·400	7·389	7·380	7·367
F	9·225	9·180	9·139	9·102	9·068	9·011

*International Union of Pure and Applied Chemistry (1978).

be surmised from prediction of the pK_a value according to the methods of Perrin *et al.* (1981), in combination with column 6 of Table 2.3. Suitable reference solutions are given in Table 2.1, most of whose pH(S) values (as they are called: S for 'standard') were derived by use of cell 11·19 of Appendix VI according to the method outlined there. The effects of variation in temperature on a selection of these standard solutions is shown in Table 2.2. The third decimal place has no significance in pH measurements, although for closely matched solutions the error can in some cases be less than 0·007 pH unit (Bates *et al.*, 1950). For more acidic standards, down even to 'pH − 12', see Table 4.2.

For the pH region above 12, special 'alkaline region' electrodes, such as the Metrohm EA 121 UX, are available. These show diminished interference from alkali metal ions, but most of them are injuriously affected by immersion in acidic solutions.

The glass electrode, when not in use should be stored by immersing the bulb in water. The calomel reference half cell is best stored by immersing its tip in 3·5 or 3·8M potassium chloride. For storage, we have found it convenient to house electrodes in flat-bottomed specimen tubes of about 2.5 cm diameter fitted with a rubber bung. A hole is bored in each bung some 2–3 mm wider than the diameter of the widest part of the electrode so that the electrode stem can easily be inserted through it. A neoprene 'O'-ring of an appropriate size is then positioned at such a height on the electrode stem that, when the electrode is passed through the bung, the tip of the electrode stays clear of the bottom of the tube. The tubes may be housed in a tube rack. The 'O'-ring also helps prevent evaporation of the liquid in the specimen tube so that the electrodes can be stored for many months. Because the electrode is not held tightly in the bung, there is no risk of the electrode adhering to the rubber and breaking upon withdrawal.

The titration is conveniently carried out in tall-form beakers (spoutless) of which 50 ml and 100 ml sizes are useful. The beaker is closed with a rubber bung bored with five holes, of which two are for the electrodes, one is for the nitrogen inlet, one for the thermometer, and one is to admit the tip of the burette

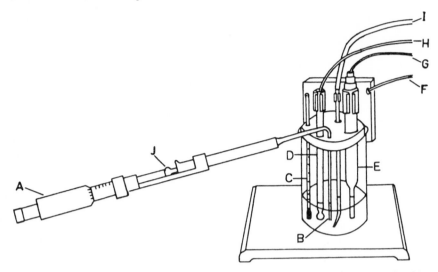

Figure 2.1 Standard titration apparatus. A, Vernier micrometer (of syringe). B, Glass needle (of syringe). C, Thermometer. D, Glass electrode. E, Calomel electrode. F, Connection to earth. G, Lead to positive terminal of pH set. H, Shielded lead to negative terminal. I, Inlet for nitrogen. J, Plunger of syringe.

or micrometer syringe. The holes for the electrodes should be of the same diameter as those bored through the bungs which position them during storage in specimen tubes (see above). When transferring the electrodes, the 'O'-rings should be slipped along the stems to ensure that the electrodes are correctly positioned in the beakers. In addition, it is advantageous, particularly when using small beakers, to secure the electrodes, with spring clips (e.g. Terry clips), to a miniature retort stand. The electrodes are connected by their leads to the terminals of the pH meter, and the stand is earthed by another wire. This arrangement is shown in Fig. 2.1.

Stirring is best accomplished by a slow stream of nitrogen bubbles. The gas is best introduced under the surface of the solution to be titrated. Its flow should be stopped during readings if it is found to interfere with them. Too fast a flow will cause loss of solution as spray.

The nitrogen should be purified by bubbling through Fieser's solution made by dissolving sodium dithionite (16 g), sodium hydroxide (15 g), and sodium anthraquinone-2-sulphonate (0·8 g) in water (100 ml). Freed from oxygen and carbon dioxide in this way, the nitrogen should then be purified from alkaline spray by bubbling through a little water.

The titrant, acid or alkali, may be delivered from a burette or from a micrometer syringe. Use of the latter, which has a vernier scale, is essential if the total volume to be delivered is small. If a burette is used for alkali, the contents can be protected from entry of carbon dioxide as shown in Fig. 2.2 (the guard tube is normally closed with a rubber bung). We use the 'Agla' micrometer syringe

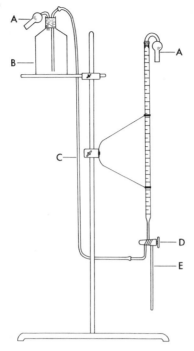

Figure 2.2 Burette for potassium hydroxide. A, Soda-lime tubes. B, $0 \cdot 1N$ potassium hydroxide in polythene bottle. C, rubber tubing. D, three-way tap. E, 15 cm stem

with the bent needles as supplied by Burroughs, Wellcome & Co. It is convenient to have two of these syringes, one for acid and one for alkali. A certain deftness is required in filling and using the alkali syringe to minimize contact with carbon dioxide from the air.

As has been explained above, good temperature control is essential for meaningful results. The room used for titrations should be free from draughts, should not vary much in temperature throughout the day, and the relative humidity should not be high. The titration vessel should stand in a bath of water which is maintained at the temperature required for the titration. The maintenance of this temperature may be either automatic or manual (e.g. by addition of hot water or chips of ice).

It is particularly important that the apparatus should be sited where no shaft of sunlight can fall on it. This precaution is to protect it from radiant heat, which can cause grossly erratic behaviour. But, for really repeatable results, it is best to work in a well-ventilated windowless room, maintained at 20 or 25°C ($\pm$ 2°C), as desired, and artificially illuminated for 24 hours a day. This is because of the photosensitivity of the silver reference cell (inside the glass electrode), illumination of which gives a difference of about 0.05 pH unit between readings made in diffuse daylight and in near-darkness (Milward, 1969; Perrin and Sayce, 1966).

2.2 Preparation of solutions

Solutions must be made in water that has been freed from carbon dioxide if alkali is the titrant. The ion-free water obtained by running distilled water through an ion-exchange column is suitable if the conductivity is sufficiently low (10^{-5} ohm^{-1}). Alternatively, 'boiled-out water' can be made by boiling distilled water vigorously for 5 min, closing the vessel with a well-cleaned rubber bung, and cooling: such water is also oxygen-free and is useful for titrating readily oxidized substances. The substances to be titrated are most conveniently dissolved by magnetic stirring, which is accomplished by inserting a small, plastic-coated iron bar in the titration vessel which is then stood on the usual magnetic flux stand. The electrodes should not be placed in the vessel until all the substance is dissolved. Magnetic stirring is best replaced by nitrogen gas stirring during the titration (see above), but the plastic-coated rod may be left in place.

If heating is used to aid dissolution, the solution must be cooled to the designated titrating temperature, or below that, before the electrodes are immersed in it. Most glass electrodes show a hysteresis effect on cooling and will not record a steady pH for some time after being cooled (Irving and Williams, 1950). There is, fortunately, no hysteresis on warming. If the temperature of the titrating room cannot be maintained at, or below, the temperature designated for titrations, the glass electrode should be stored in a thermostat, at the designated temperature.

The alkali used for titrant must be carbonate-free, as judged from a barium hydroxide test or by potentiometric titration of histidine (see below). We have found that the volumetric solutions of commerce are usually carbonate-free. The choice lies between sodium and potassium hydroxide, but there is less electrode-error with the latter at high pH values. For most purposes, 0·1N is a suitable strength. In the following paragraphs we describe the preparation of carbonate-free 0·1N potassium hydroxide to help any one who has to prepare his own. Others may proceed to Section 2.3.

The commercial pellets of potassium hydroxide usually carry all the carbonate on the outside. Hence it is possible to wash a weighed amount of potassium hydroxide (the fused, 15% hydrated solid), to titrate these washings (which are then rejected), and hence find how much water to add to the residue to make slightly stronger than decinormal potassium hydroxide. A further titration and dilution should give exactly 0·100N KOH. These operations should be conducted in some easily contrived apparatus entirely out of contact with the carbon dioxide of the air. The absence of carbonate is usually confirmed by potentiometric titration of the amino-acid histidine, for whose pK_a of 6·08 no agreeing set of nine values can be obtained in the presence of a disturbing amount of carbon dioxide. Should the above method fail, through the specimen of potassium hydroxide having carbonate-bearing fissures throughout, the following ion-exchange method is strongly recommended. The procedure falls into three sections, denoted (a), (b) and (c).

In summary, barium hydroxide is added to a solution of commercial potassium hydroxide (analytical grade). The precipitate of barium carbonate is allowed to settle and the excess barium ions removed by passing the solutions through a column of ion-exchange resin (Amberlite IR 120) which is quantitatively in the potassium form: RSO_3K (R is the matrix of the resin). The eluate is a solution of pure potassium hydroxide free from carbonate.

(a) Precipitation of carbonate: A.R. potassium hydroxide (14 g) is dissolved in ion-free water (about 1·5 litres), and A.R. barium hydroxide (3 g) is added. The suspension is shaken for 15 min in a conical flask sealed with a rubber stopper through which a glass tube protrudes, by 5 cm, into the flask. The flask is inverted and allowed to stand overnight so that the precipitate of barium carbonate settles below the level of the outlet tube

(b) Preparation of the resin: Amberlite resin IR 120H (50 ml analytical grade) is made into a slurry with water and poured into a column fitted with a three-way double oblique bore stopcock. The resin is backwashed with N HCl (1 litre) and then with water (*ca.* 1·5 litres) to remove excess acid. Next, 0·2N potassium chloride solution (about 1·5 litres) is passed through the column from the top until the pH values of the entrant and eluate solutions are identical (about 1·3 litres are required). The potassium chloride solution is then allowed to remain in contact with the resin for a further hour and the remaining 200 ml is then passed through. This ensures that the resin is quantitatively in the potassium form. The excess potassium chloride solution is removed by downwashing the column with ion-free water until 100 ml of eluate gives no turbidity with acidified silver nitrate solution (approximately 1·8 litres water is required). To ensure that the water flowing through the resin is carbonate-free, the washing is carried out in an inert atmosphere using the arrangement shown in Fig. 2.3. The wash water is discarded.

(c) Purification of the carbonate-free potassium hydroxide solution containing barium ions: the conical flask containing the impure potassium hydroxide is connected to the top of the column and the solution passed through the column. The first 100 ml is rejected and the remainder passed through the polythene tube into a nitrogen-filled 2-litre polythene bottle previously calibrated in 100 ml divisions and fitted with a soda-lime guard tube. When the flow from the conical flask ceases, a calibrated 500-ml polythene bottle fitted with a polythene syphon tube and soda-lime guard tube and containing ion-free water is attached to the top of the column. The resin is then washed with water (200 ml) and the eluate is combined with the potassium hydroxide solution in the polythene bottle. The tube connecting the column to the polythene bottle is sealed with a screw-clip and then disconnected from the column. Nitrogen is bubbled through the potassium hydroxide solution for 10 min with occasional swirling to ensure a homogeneous solution. The bottle is then connected to a 50-ml burette and the potassium hydroxide solution is standardized against A.R. potassium hydrogen phthalate (dried 1 h at 120°C) using phenolphthalein as indicator. The volume in the bottle is adjusted by further small additions of

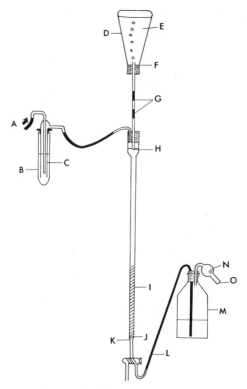

Figure 2.3 Preparation of carbonate-free potassium hydroxide. A, Inlet for nitrogen. B, Pressure regulator, Pregl's type with loosely-fitting bung (Gallenkamp Cat. No. 13607/1). C, N Potassium hydroxide (as CO_2 trap). D, Conical flask (2 litre). E, Potassium and barium hydroxides. F, Precipated barium carbonate. G, Polythene tubing, suitable for clamping. H, End ground to 45°. I, Amberlite resin (IR 120) as potassium form. J, Glass bead. K, Slight constriction. L, Polythene tubing to base of polythene bottle. M, Polythene bottle (2·2 litres capacity). N, Soda lime. O, Cotton-plug.

ion-free water until the potassium hydroxide solution is exactly 0·1000N, the normality being checked after each addition. The yield is about 2 litres. We find that the carbonate content, determined on 800 ml by a standardized micro-gravimetric method (the increase in weight of soda-lime after the liberation of CO_2 by excess acid), is less than 1 ppm.

2.3 Choice of concentration for the titration

We recommend that substances be titrated at 0.01M concentration wherever solubility permits. At this concentration, activity effects are usually small (see p. 50). However, concentrations up to 0.1M may be used if activity corrections, as on p. 51, are made.

It must now be asked, what lower and upper limits are imposed on the accurate determination of pK_a by the method of this chapter. A lower limit is imposed by the uncertainty of applying activity coefficients in the pH range 1–2 (the magnitude of the effect depends on the concentration and the pH values observed during the titration). We place the lower limit for accuracy at pK_a 2.5, but reasonable estimates of pK_a in the range 1.2–2.5 can be obtained provided activity corrections are applied. The upper limit is imposed by the unreliable performance of the glass electrode at high pH; we somewhat arbitrarily set pK_a 11 as the limit. Above this value, the application of activity coefficients can also become uncertain. However, using a different type of approach, a pK_a value as high as 13·65 was determined by the hydrogen electrode (Hall and Sprinkle, 1932; Thamsen, 1952).

Misleading estimates can be expected if the pK_a is less than the negative logarithm of the concentration. Thus 0·01M is too dilute for extracting any pK_a less than 2.

Because of poor solubility, it may be desired to titrate a solution more dilute than 0.01M. The above rule of the negative logarithm of the dilution will indicate, after a preliminary titration, if reliable results could be expected at such a dilution. Where doubt exists, the pH values arising during the titration of a similar volume of water should be compared with the pH values arising during the dubious titration, to see if the two series of figures differ significantly.

The ionic strength of a 0·01M solution of a univalent substance, when half neutralized, is 0·005. Some workers like to titrate all solutions 'at constant ionic strength', usually by making them 0.1M in potassium chloride. However, this should not be done when using the methods described in this chapter which avoid activity calculations (they also avoid corrections for liquid junction potentials). As an indication of the degree of accuracy obtainable, 0.01M acetic acid was titrated at 20°C and gave pK_a 4.74 $\pm$ 0·03. This can be compared with the thermodynamic pK_a (4·7560) which was obtained by using every possible refinement of technique and calculations (Harned and Ehlers, 1932).

2.4 Details of the titration method

The temperature dial on the pH apparatus is to be set to the required temperature. Next, the glass electrode is standardized with two buffers as described on p. 17, at a concentration suggested by the considerations discussed on p. 24. The solution is brought to the required temperature, and stirring by a slow stream of nitrogen (see p. 22) is begun. The pH is read. The titrant is then added in ten equal portions, each a tenth of an equivalent, and the pH is recorded as soon as equilibrium is reached after each addition. No credence can be given to pKs which are reported as 'the pH recorded at half neutralization'.

When the titration is complete, the electrodes are at once washed, and placed in either phthalate, borate, or phosphate buffer (choosing the buffer which is nearest to the pK found). If its pH is not reproduced to $\pm$ 0.02 *without adjustment of the set*, the titration results must be discarded. If the potential repeatedly strays in this way during a titration, one of the electrodes may be faulty, or the temperature may not be as steady as is supposed.

Dilution of the solution by the titrant causes little error provided (a) that the titrant is at least ten times as concentrated as the substance being titrated and (b) that the concentration on which calculations are based is reached at the mid-point of the titration as in Tables 2.3 and 2.4. By using a titrant which is 100 times as concentrated as the substance, the latter can be prepared for titration in the concentration on which calculations are based, as in Table 2.5. If, however, it is desired to correct for the dilution caused by the water introduced by the titrant, an adjustment can be made as follows.

Insert, into the table, an extra column giving the total concentration of material being titrated, regardless of species. For example in Table 2.4, this column (2a) would come between columns 2 and 3, and the figures in it would progressively decline according to the progressive dilution caused by the titrant, passing through 0·01 at 2·5 ml. The figures for columns 3 and 4 should next be adjusted so that, on every line, they add up to the value found in column 2a. The rest of the table remains as before.

The most difficult titrations with the glass electrode are those of substances with pK_a near to 11. The following sources of error can contribute to an inaccurate result:

(a) Standardization of the electrode at a pH no higher than that of borate buffer (about pH 9.2) which is too far from the pH region covered in the titration;
(b) The nature of the glass electrode which becomes more and more responsive to potassium ions as the ratio K^+/H^+ increases;
(c) A minute but significant amount of carbon dioxide may enter from the air even when great care is taken to exclude it.

Hence we recommend that potentiometry with the glass electrode be not used for substances of pK_a more than 11; nor, above pH 11, for solutions more dilute than 0.01M. Spectrophotometry (Chapter 4) is more suited for the high pK_a region; should this not be practicable, recourse may be had to thermometric titration (Chapter 7), conductimetry (Chapter 6), or potentiometry with the hydrogen electrode (Chapter 3).

2.5 Derivation and choice of equations for calculating pK_a

If the pH remains between 4 and 10 when a 0·01M solution is being titrated, only the very simple calculations shown in Table 2.3 (p. 28) need be carried out. The figures in column 6 occupy a key position, being a numerical expression of the logarithmic term in equation (1.5a) (p. 5). These figures (column 6) are independent of the dilution, few in number, and very useful to commit to memory. Thus, if titration conditions permit the use of this simplified procedure, only columns 1, 2, 6 and 7 need be written in the practical notebook, and of these, column 6 is always the same.

Before tackling the slightly more complicated examples where hydrogen ions or hydroxyl ions, derived from hydrolysis of the salt being formed, are present in interfering amounts, it would be well to look into the nature of entries in the first five columns of Table 2.3. Column one is simply derived from the burette readings. Column 2 contains the corresponding readings on the pH set.

In column 3, the top line of figures reads 0·010, which is C_0, namely the original concentration of substance undergoing titration. We diminish this figure by one-tenth on each succeding line, to correspond to each tenth equivalent of alkali added. In the present simple case, this column could equally properly be headed $C_0 - [KOH]$, interpreting $[KOH]$ as the concentration that the alkali would assume if no boric acid were present. It would also be proper to label it quite succinctly $[HA]$, namely the concentration of non-ionized material at each point of the titration. But this expression is short for $[HA] + [OH^-] - [H^+]$ the last two terms being insignificant in the case of boric acid but highly significant in the following tables. Hence, we have come to prefer the heading as printed, because it applies to all cases and is sufficiently brief.

Column 4, headed C_0 – column 3, is, in the present case, simply the concentration that the potassium hydroxide would assume in the absence of boric acid. It is also the same as $[A^-]$, the concentration of the anion being produced by the titration. However, in the following tables it would become $[A^-] - [OH^-] + [H^+]$, so it has been thought better to conserve the universal heading.

Column 5 contains the quotients obtained by dividing figures in column 3 by the corresponding ones in column 4.

We are now ready to tackle the more complicated cases in which the pH range of a 0·01M titration falls outside the limits pH 4–10. Even in Table 2.3, the last usable reading (pH 10·14) falls outside this range. If this value is corrected for hydroxyl ion concentration as in Table 2.6, the corresponding pK_a value rises from 9·19 to 9·23, and the final result is sharpened from 9·25 ± 0·06 to 9·26 ± 0·03.

If the solution being titrated is more dilute, the pH must remain between narrower values throughout the titration if such corrections are to be avoided. Thus for a 0·001M solution, the pH must remain between 5 and 9.

The necessity for allowing for hydrogen ion, and/or hydroxyl ion, concentrations will now be explained in more detail. Because all solutions are *electrically* neutral, the sum of all positive charges must equal the sum of all negative charges, thus,

$$[A^-] + [OH^-] = [K^+] + [H^+]. \tag{2.4a}$$

Because all salts are completely ionized, $[K^+]$ equals the concentration of potassium hydroxide (after allowing for dilution by the solution). Hence

$$[A^-] = [KOH] - [OH^-] + [H^+]. \tag{2.4b}$$

$[C_0]$, the total concentration of acid taken, is present in two forms, A^- and HA. Consequently,

$$[C_0] = [A^-] + [HA]. \tag{2.4c}$$

By combining (2.4b and c), we get:

$$[HA] = [C_0] - [KOH] + [OH^-] - [H^+]. \tag{2.4d}$$

In these equations, $[KOH]$ represents the concentration that the alkali would achieve by dilution if no other substance were present in the solution. The expression $[HCl]$, on p. 32, is used similarly.

Thus $[A^-]$ is $[KOH] - [OH^-]$ in the alkaline region, from equation (2.4b) whereas it is $[KOH] + [H^+]$ in the acidic region. Similarly, $[HA]$ is $[C_0] - [KOH] + [OH^-]$ in the alkaline region, and so on. This modification of the calculations to include the contribution of hydroxyl ions and hydrogen ions refines the results most gratifyingly. How to calculate $[H^+]$ and $[OH^-]$ from the pH reading is set out in Table 11.3 of Appendix III. However, we prefer to use

Table 2.3   Determination of the ionization constant of a monobasic acid requiring no correction for hydrogen or hydroxyl ions

Substance: Boric acid. $H_3BO_3 = 61\cdot84$. *Temperature:* 20°C.
Concentration: 0·01M at half neutralization. Boric acid (0·0309 g), dried overnight in a vacuum desiccator ($CaCl_2$, 20 mm, 20°C), was dissolved in 47·5 ml boiled-out water.

1	2	3	4	5	6	7
Titrant 0·1N KOH (ml)	Meter reading (pH)	C_0 (the original conc.) diminished by tenths	C_0 minus column 3	Column 3 divided by column 4	log of column 5	pK_a (columns 2 + 6)
0	6·16	0·010	0			
0·5	8·34	0·009	0·001	9/1	0·95	9·29
1·0	8·68	0·008	0·002	8/2	0·60	9·28
1·5	8·89	0·007	0·003	7/3	0·37	9·26
2·0	9·07	0·006	0·004	6/4	0·18	9·25
2·5	9·26	0·005	0·005	5/5	0	9·26
3·0	9·43	0·004	0·006	4/6	−0·18	9·25
3·5	9·62	0·003	0·007	3/7	−0·37	9·25
4·0	9·84	0·002	0·008	2/8	−0·60	9·24
4·5	10·14	0·001	0·009	1/9	−0·95	9·19
5·0	10·56	0	0·010			

Result: $pK_a = 9\cdot25$ ($\pm0\cdot06$) at 0·01M and 20°C (using all nine values in the set). If the pH 10·14 reading is corrected for $[OH^-]$, as in Table 2.6, the value 9·19 becomes 9·23 and the final result is sharpened to $9\cdot26 \pm 0\cdot03$.

Table 2.4   Determination of the ionization constant of a monobasic acid requiring correction for the concentration of hydrogen ions

Substance: Benzoic acid. $C_7H_6O_2 = 122.1$. *Temperature:* 20° C.
Concentration: 0.01M at half-neutralization. Benzoic acid (0.0611 g), dried overnight in a vacuum desiccator (H_2SO_4, 20 mm, 20°C), was dissolved in 47.5 ml water by magnetically stirring for 20 min at 40° C.

1	2	3	4	5	6	7	8
Titrant 0.1N KOH (ml)	pH	C_0 diminished by tenths	C_0 minus column 3	$\{H^+\}$*	Columns (3 − 5) divided by columns (4 + 5)	log of column 6	pK_a (columns 2 + 7)
0.0	3.10	0.010	0				
0.5	3.38	0.009	0.001	0.000 42	858/142	+ 0.78	4.16
1.0	3.63	0.008	0.002	0.000 23	777/223	+ 0.54	4.17
1.5	3.83	0.007	0.003	0.000 15	685/315	+ 0.34	4.17
2.0	3.99	0.006	0.004	0.000 10	590/410	+ 0.16	4.15
2.5	4.16	0.005	0.005	0.000 07	493/507	− 0.01	4.15
3.0	4.34	0.004	0.006	0.000 05	395/605	− 0.18	4.16
3.5	4.53	0.003	0.007	0.000 03	297/703	− 0.37	4.16
4.0	4.76	0.002	0.008	−	200/800	− 0.60	4.16
4.5	5.12	0.001	0.009	−	100/900	− 0.95	4.17
5.0	8.18	0	0.010	−			

Result: $pK_a = 4.16$ (± 0.01) at 0.01M and 20°C, (using all nine values in the set)
 *From column 2 and Table 11.3 (p. 198).

$\{H^+\}$ and $\{OH^-\}$, namely the corresponding activities, to achieve a *mixed constant* (K_a^M), rather than a concentration constant K_a^C, so as to be closer to the thermodynamic constant (K_a^T), as explained on p. 47.

So far, we have dealt with the titrations of acids only, but the ionization constant of a base can be obtained similarly, from equation (1.6a). When the titration lies between pH 4 and 10 (for a 0.01M solution), hydrogen ion and hydroxyl ion corrections can be neglected, as in Table 2.7. Above pH 10, the hydroxyl ion activity should be incorporated as in Table 2.9, and below pH 4, the hydrogen ion activity must also be used in the calculations, as in Table 2.8.

2.6 Some typical titrations (worked examples)

These worked examples cover the commoner types of potentiometric titration and show how the results are best tabulated.

Table 2.3 shows the titration of 0.01M boric acid, and is typical of a titration

Table 2.5   Determination of the ionization constant of a monobasic acid in the form of its sodium salt (cf. the titration of the free acid in Table 2.4)

Substance: Sodium benzoate $C_7H_5O_2Na = 144.1$. *Temperature:* 20°C.
Concentration: 0.01M throughout. Sodium benzoate (0.0721 g), dried overnight in a vacuum desiccator (H_2SO_4, 20 mm, 20°C), was dissolved in 50 ml cold water.

1	2	3	4	5	6	7	8
Titrant N HCl (ml)	pH	C_0 minus column 4	C_0 diminished by tenths	$\{H^+\}$	Columns $(3-5)$ divided by columns $(4+5)$	log of column 6	pK_a (columns $2+7$)
0.00	8.12	0	0.010				
0.05	5.09	0.001	0.009	–	100/900	– 0.95	4.14
0.10	4.75	0.002	0.008	–	200/800	– 0.60	4.15
0.15	4.53	0.003	0.007	0.000 03	297/703	– 0.37	4.16
0.20	4.34	0.004	0.006	0.000 05	395/605	– 0.18	4.16
0.25	4.16	0.005	0.005	0.000 07	493/507	– 0.01	4.15
0.30	3.99	0.006	0.004	0.000 10	590/410	+ 0.16	4.15
0.35	3.82	0.007	0.003	0.000 15	685/315	+ 0.34	4.16
0.40	3.62	0.008	0.002	0.000 24	776/224	+ 0.54	4.16
0.45	3.37	0.009	0.001	0.000 43	857/143	+ 0.78	4.15
0.50	3.09	0.010	0				

Result: $pK_a = 4.15$ (± 0.01) at 0.01M and 20°C (using all nine values in the set).

not requiring hydrogen or hydroxyl ion corrections (see above). Actually the last value in the set does rise above pH 10 and the pK_a calculated from it is improved by a correction as explained on p. 27.

Table 2.4 (benzoic acid) is typical of a titration requiring correction for *hydrogen* ions. Table 2.5, the titration of 0.01M sodium benzoate with strong acid, is virtually the same titration as in Table 2.4, but approached from the opposite direction. Comparison of the pair of Tables 2.4 and 2.5 will show how any titration can be tackled from alternative directions without substantial changes in the calculations. Table 2.6, the titration of *p*-cresol, provides an example of an acid requiring correction for *hydroxyl* ions.

In Table 2.7 the much used buffer 'Tris' is titrated as an example of a base not requiring correction. Table 2.8 gives an example (*p*-chloroaniline) of a base requiring correction for *hydrogen* ions. Table 2.9 shows the titration of glycine, from the zwitterion $H_3N^+ \cdot CH_2CO_2^-$ to the anion $H_2N \cdot CH_2CO_2^-$. Zwitterions will be discussed in Chapter 8, and it may be sufficient now to say that Table 2.9 shows the titration of an amino group from its cation to the neutral form. This example illustrates correction for *hydroxyl* ions in the titration of a base.

Table 2.6   Determination of the ionization constant of a monobasic acid requiring correction for the concentration of hydroxyl ions

Substance: p-Cresol, $C_7H_8O = 108.13$. *Temperature:* 20°C.
Concentration: 0.01M at half-neutralization. p-Cresol (0.0541 g), purified by vacuum distillation, was ground and dried overnight in a vacuum desiccator (H_2SO_4, 20 mm, 20°C), and dissolved in 47.5 ml ion-free water by magnetically stirring for 30 min at 20°C under nitrogen.

1	2	3	4	5	6	7	8
Titrant 0.1N KOH (ml)	pH	C_0 diminished by tenths	C_0 minus column 3	$\{OH^-\}$*	Columns (3 + 5) divided by columns (4 − 5)	log of column 6	pK_a (columns 2 + 7)
0	6.92	0.010	0	–			
0.5	9.19	0.009	0.001	–	900/100	+ 0.95	10.14
1.0	9.55	0.008	0.002	0.000 02	802/198	+ 0.61	10.16
1.5	9.77	0.007	0.003	0.000 04	704/296	+ 0.38	10.15
2.0	9.97	0.006	0.004	0.000 06	606/394	+ 0.19	10.16
2.5	10.14	0.005	0.005	0.000 10	510/490	+ 0.02	10.16
3.0	10.29	0.004	0.006	0.000 13	413/587	− 0.15	10.14
3.5	10.46	0.003	0.007	0.000 20	320/680	− 0.33	10.13
4.0	10.64	0.002	0.008	0.000 30	230/770	− 0.53	10.11
4.5	10.84	0.001	0.009	0.000 48	148/852	− 0.76	10.08
5.0	11.08	0	0.010				

Result: $pK_a = 10.14$ (± 0.03) at 0.01M and 20°C (using the inner seven values of the set).
  *From column 2 and Table 11.3 (p. 198).

2.7 Precision and accuracy. Checking the precision obtained

Although a result can be precise (i.e. contained within a narrow spread) without being accurate (i.e. true), the reverse is not the case. Hence it is desirable to see that the average of all values, in a set of nine values, falls within a spread of ±0.06 (see p. 10). With care and experience, this scatter can be greatly reduced as is evident from Tables 2.4 to 2.9 which are records of routine titrations in our laboratories. The commonest causes of too large a spread are (a) the presence of impurities in the substance being titrated, and (b) small inaccuracies in adding the titrant (see below under 'common sources of error'). Such errors often are most evident in the first or ninth value of a set, leaving the inner seven values acceptably close.

2.8 Common sources of error, and their elimination

One of the commonest errors in titrating with alkali is for the values, in the set of pK_a values, to show an *upward trend* as the titration progresses. This is usually

Table 2.7   Determination of the ionization constant of a monoacidic base requiring no correction for hydrogen or hydroxyl ions

Substance: Aminotrishydroxymethylmethane   (i.e.   the   well-known   buffer,   'Tris').
$H_2N \cdot C(CH_2OH)_3 = C_4H_{11}NO_3 = 121 \cdot 14.$ *Temperature:* 20°C.
Concentration: 0·01M throughout. Tris (0.0606 g), dried for 1 h in air at 110°C, was dissolved in 50 ml ion-free water and titrated under nitrogen (to exclude carbon dioxide).

1	2	3	4	5	6	7
Titrant N HCl (ml)	pH	C_0 minus column 4	C_0 diminished by tenths	Column 3 divided by column 4	log of column 5	pK_a (columns 2 + 6)
0	10·12	0	0·010			
0·05	9·12	0·001	0·009	1/9	− 0·95	8·17
0·10	8·78	0·002	0·008	2/8	− 0·60	8·18
0·15	8·55	0·003	0·007	3/7	− 0·37	8·18
0·20	8·36	0·004	0·006	4/6	− 0·18	8·18
0·25	8·19	0·005	0·005	5/5	0·00	8·19
0·30	8·01	0·006	0·004	6/4	+ 0·18	8·19
0·35	7·81	0·007	0·003	7/3	+ 0·37	8·18
0·40	7·57	0·008	0·002	8/2	+ 0·60	8·17
0·45	7·21	0·009	0·001	9/1	+ 0·95	8·16
0·50	4·32	0·010	0			

Results: $pK_a = 8 \cdot 18 \, (\pm 0 \cdot 02)$ at 0·01M and 20°C (using all nine values in the set).

caused by an impurity in the substance undergoing determination, so that not so much of it is present as had been supposed. By far the commonest and most troublesome impurity is water. To avoid this trouble, every substance submitted for determination of pK_a should be of analytical purity and dried under the same conditions that preceded its analysis. When this is not possible, as with a deliquescent substance, one can make use of the function Z defined, at any part of the titration, as the sum of four known concentrations:

$$Z = [HCl] - [KOH] - [H^+] + [OH^-].$$

Benet and Goyan (1967) now write (for an acid):

$$Z = C_0 - (1/K_a^C)Z[H^+],$$

and plot Z against $Z[H^+]$. This gives a straight line with a slope equal to the negative reciprocal of the ionization constant, and an intercept equal to C_0. The corresponding equation for bases is:

$$Z = C_0 - K_a^C(Z/[H^+]),$$

which allows a plot of Z against $Z/[H^+]$ as a straight line with a slope equal to the ionization constant bearing a negative sign, and an intercept of C_0.

Table 2.8 Determination of the ionization constant of a monoacidic base requiring correction for the concentration of hydrogen ions

Substance: p-Chloroaniline ($C_6H_6NCl = 127.57$). *Temperature:* 20°C.
Concentration: 0·01M throughout. p-Chloroaniline (0·0638 g), purified by vacuum distillation, ground and dried overnight in a vacuum desiccator ($CaCl_2$, 20 mm, 20°C), was dissolved in 50 ml water by magnetically stirring at 40°C for 30 min.

1	2	3	4	5	6	7	8
Titrant N HCl (ml)	pH	C_0 minus column 4	C_0 diminished by tenths	$\{H^+\}$*	Columns (3 − 5) divided by (4 + 5)	log of column 6	pK_a (columns 2 + 7)
0	6·72	0	0·010				
0·05	4·85	0·001	0·009	−	100/900	− 0·95	3·90
0·10	4·52	0·002	0·008	0·000 03	197/803	− 0·61	3·91
0·15	4·31	0·003	0·007	0·000 05	295/705	− 0·38	3·93
0·20	4·14	0·004	0·006	0·000 07	393/607	− 0·19	3·95
0·25	3·96	0·005	0·005	0·000 11	489/511	− 0·02	3·94
0·30	3·81	0·006	0·004	0·000 15	585/415	+ 0·15	3·96
0·35	3·64	0·007	0·003	0·000 23	677/323	+ 0·32	3·96
0·40	3·43	0·008	0·002	0·000 37	763/237	+ 0·51	3·94
0·45	3·20	0·009	0·001	0·000 63	837/163	+ 0·71	3·91
0·50	2·95	0·010	0				

Result: pK_a = 3·93 (± 0·03) at 0·01M and 20°C (using all nine values in the set).
 *From column 2 and Table 11.3 (p. 198)

Another possible cause of an upward pH trend is that the stream of nitrogen is too fast and is expelling some of the solution as spray. Another cause of an upward trend is that the correct amount of material is present throughout the titration but not all of it is in solution. In the presence of undissolved material, no satisfactory pK_a value can be reached.

When titrating with acid, a *downward trend* can be due to the same causes as given in the last paragraph for an upward trend with alkali.

Another common source of error is that the titrant is not added in exactly the desired volume. Sometimes this is due to inexperience in manipulation, but sometimes the burette is in error.

Occasionally the required degree of precision is not obtained because the method is not suitable for the substance (pp. 24–25 may be re-read in this connection).

An incongruous set of results often indicates that decomposition is occurring during titration. Sometimes the first few readings give concordant pKs after which decomposition causes a drift of potential. Those substances which are easily decomposed by acid or alkali are often attacked by each drop of titrant,

Table 2.9   Determination of the ionization constant of a monoacidic base requiring correction for the concentration of hydroxyl ions

Substance: Glycine. $C_2H_5NO_2 = 75.07$. *Temperature:* 20°C.
Concentration: 0·01M at half-neutralization. Glycine (0·0375 g), recrystallized from water and dried in air at 110°C for 1 h, was dissolved in 47·5 ml ion-free water and titrated under nitrogen (to exclude carbon dioxide).

1	2	3	4	5	6	7	8
Titrant 0·1N KOH	pH	C_0 diminished by tenths	C_0 minus column 3	$\{OH^-\}$*	Columns 3 + 5 divided by columns 4 − 5	log of column 6	pK_a (columns 2 + 7)
0	6·30	0·010	0	—			
0·5	8·94	0·009	0·001	—	900/100	+ 0·95	9·89
1·0	9·28	0·008	0·002	—	800/200	+ 0·60	9·88
1·5	9·50	0·007	0·003	0·000 02	702/298	+ 0·37	9·87
2·0	9·69	0·006	0·004	0·000 03	603/397	+ 0·18	9·87
2·5	9·88	0·005	0·005	0·000 05	505/495	+ 0·01	9·89
3·0	10·05	0·004	0·006	0·000 08	408/592	− 0·16	9·89
3·5	10·23	0·003	0·007	0·000 15	315/685	− 0·34	9·89
4·0	10·42	0·002	0·008	0·000 18	218/782	− 0·55	9·87
4·5	10·68	0·001	0·009	0·000 33	133/867	− 0·81	9·87
5·0	11·01	0	0·010				

Results: $pK_a = 9.88$ (± 0.01) at 0·01M and 20°C (using all nine values in the set).
  *From column 2, and Table 11.3 (p. 198)

even in well stirred solutions. Sometimes back-titration gives a different set of pH values from the forward titration, whereupon a second forward titration retraces the pH values of the first (i.e. hysteresis loop is formed). This indicates a slow and quantitative interconversion of two related substances, which may be tautomers, or they may have a ring-chain relationship, or one may be a covalent hydrate of the other.* Pseudo-acids (e.g. nitromethane) and pseudo-bases (e.g. triphenylmethane dyes and quaternary *N*-heterocycles) also show this phenomenon and equilibration may take anything from a minute to a day. Suitable methods exist for tackling covalent hydration (Albert and Armarego, 1965; Perrin, 1965b), and pseudo-base formation (Goldacre and Phillips, 1949); normal mass action relationships are followed, and equilibrium constants can be extracted from the data, quite apart from any study of the kinetics of the reaction.

*For reviews of covalent hydration, see Albert (1967, 1976); for hydrated and anhydrous forms of pyruvic acid, see Table 9.1.

Ill-defined acidic or basic constants are often obtained at low pH values, especially below pH 2. In some cases the substance really has acidic or basic properties, as suggested by the pK_a, but the result has too great a spread to be acceptable. These results arise when the pK_a is equal to, or slightly above, the logarithm of the dilution. In one base which we investigated (the concentration was 0·05M and hence the dilution was 20, and the log dilution 1·3), the pK_a seemed to be 1·45 ±0·13 (volume corrections for the added titrant had been made as on p. 26). Such a large spread is inadmissible, and the error was traced to the use of activities and concentrations in the same equation, i.e. $\{H^+\}$ had been calculated as antilog $(0 - pH)$, and this *activity* term had been combined with stoichiometric *concentration* terms for $[B]$ and $[BH^+]$. When the concentration term $[BH^+]$ was converted to an activity term, as on pp. 51–52, a satisfactory thermodynamic pK_a of 1·25 ± 0·05 was obtained (see p. 48). The use of $\{H^+\}$ with concentration terms does not give trouble if the pK_a is 1·5 units, or more, above the logarithm of the dilution.

2.9 False constants

When a pK_a appears, in a titration, to be well below the logarithm of the dilution at which it was found, it is almost certainly a mirage (false constant) and the substance being titrated has no acidic or basic properties in that region. Thus the titration of pure water (4·75 ml), with 0·5 ml N HCl, gives a pK_a of 0·62 ± 0·72 (even if correction is made for dilution by the titrant), so long as no activity correction is performed. This result is clearly absurd. Yet, had an inert substance been dissolved in the water (say at 0·1M), it may have been credited with this pK_a, although the fact that the logarithm of the dilution is 1·0 should put the investigator on his guard.

False acidic or basic constants can also be found in glass-electrode titrations when most of the pH readings are higher than 11. For example, 100 ml boiled-out water, titrated with 5·0 ml, 0·1N potassium hydroxide, gave a false pK_a of 12·11 ± 0·05 (calculations based as on a 0·005M solution). Had 0·005M of an inert substance been present, it might have been credited with this constant. For this reason, all pK_a values in the literature, if outside the range 2–11, should be scrutinized carefully.

2.10 Partly aqueous solvents

When a substance is poorly soluble in water, but highly soluble in a volatile solvent, it is natural to consider determining the ionization constant in a mixture of the two solvents, e.g. in a 1:1 mixture of water and ethanol. Such a course of action has pitfalls, as will become apparent in what follows.

The practice stems largely from a series of papers by Mizutani (1925) on 'The Dissociation of Weak Electrolytes in Dilute Alcohols'. It was found, as would be expected, that alcohols weaken both acids and bases: e.g. the pK_a of an acid was raised by about 1·0, and that of a base lowered by about 0·5 (maximum 0·89, minimum 0·30) in 60% methanol. Hall and Sprinkle (1932) plotted the curves of pK_a against decreasing alcohol concentration from 97% to 10% ethanol for 18 aliphatic and aromatic amines. Various amines had different slopes, but in some cases satisfactory extrapolation to 0% alcohol was possible. They found that the average depression of pK_a by 50% ethanol was 0·54 (maximum 0·88, minimum 0·26). The extrapolation of these 'hockey-stick' shaped curves proved impossible when the amine was too insoluble to give values in 10% and 20% ethanol, because successful extrapolation depended very much on knowing these values.

Titrations in 50% acetone have revealed still greater depressions of pK_a (1·5 to 2·5 units) (Pring, 1924), and in 50% dioxan they are often greater still. Yet, in one instance, a series of curves was found more regular, and easier to extrapolate, in 50% acetone than in dilute alcohols (Cavill *et al.*, 1949).

It has often been said that comparison of the strength of a series of substances in a partly aqueous solvent is valid if the substances are chemically related. That this need not be true is illustrated by Table 2.10. There it can be seen that, in aqueous solvents of ever-increasing ethanolic content, aniline and its *N*-methylated derivatives steadily become weaker bases. Yet the effect of the ethanol is least on the non-methylated substance, with this paradoxical result: methylation seems to increase basic strength (in 0–35% ethanol) yet also to decrease it (in 50–65% ethanol). Mixtures of methanol and water create similar anomalies with the pK_a values of aliphatic amines and also azo-dyes (de Ligny *et al.*, 1961). Dilute dimethylformamide, and 2-methoxyethanol (methyl cellosolve) have been found similarly troublesome.

Table 2.10 pK_a values of some chemically related amines in dilute ethanol (25°C)

	Ethanol (% by weight)				
	0	20	35	50	65
Aniline	4·64	4·42	4·16	3·92	3·80
Methylaniline	4·84	4·62	4·28	3·90	3·64
Dimethylaniline	5·01	4·75	4·30	3·81	3·50

Gutbezahl and Grunwald (1953)

Such irregularities are often encountered when two chemically related substances differ in liposolubility. This follows from the dependence of ΔpK_a (the difference between the pK_a of a substance in water and its pK_a in a partly aqueous solvent) on the distribution coefficients of the various species involved in the equilibrium (Kolthoff *et al.*, 1938), thus:

$$\Delta pK_a = \log D_{H+} + \log D_B - \log D_{BH+}, \qquad (2.5)$$

where D_X is the distribution coefficient of X between the two solvents (the more lipophilic the species, the higher the coefficient).

Thus the more lipophilic species can be surrounded by a cage of solvent molecules of low dielectric constant, while the ion is surrounded mainly by water molecules. When B is highly lipophilic and BH^+ is not, a large ΔpK_a can be expected. This value should pass through a maximum with increasing lipophilic nature because eventually even the ion must be more soluble in the organic solvent than in water. Such a maximum is seen in the series of benzologues: pyridine, acridine and 3, 4-benzacridine where ΔpK_a is respectively 0·73, 1·49, and 0·54 (water–50% alcohol) (Albert, 1966a).

Acids behave similarly. Thus, in water benzoic acid is 4 times as strong as acetic acid, but in 20% ethanol only 2·5 times as strong, and in 50% ethanol they have the same strength (Grunwald and Berkowitz, 1951).

It is becoming clear that what keeps the aqueous and partly aqueous pK values apart is not only the rejection of hydration by the molecule in proportion to its lipophilicity, but also some steric hindrance to hydration in proportion to geometrical protection of the molecule's ionizing area. When factors for both of these effects have been worked out and reduced, let us hope, to an easily-applied formula, it should be possible to convert results obtained in mixed solvents to their desired values in water. Let us not, however, underestimate the difficulties of bringing this about.

In their five pages-long review of this subject, Benet and Goyan (1967) conclude: 'As has been reviewed above, there is presently no completely satisfactory method for converting p_sK_a values (i.e. those obtained in partly aqueous solutions) to pK_a values. Therefore, although good accuracy can be realized in determining p_sK_a values, this accuracy cannot be carried over to the aqueous dissociation constants.' They go on to recommend aqueous methanol as the least error-prone of mixed solvents because of a greater accumulation of information about the behaviour of the glass electrode in this solvent and about thermodynamic effects of steadily diminishing the content of methanol in the mixture. To avoid extrapolation of the hockey-stick shaped inflection (see above), these authors recommend extrapolation from the *linear* plot of $p_sK_a + \log[H_2O]$ against $1/D$ (where D is the dielectric constant), a procedure due to Shedlovsky (1962). For this approach to succeed, Benet and Goyan recommend that the methanol content does not exceed 40%.

The use of partly aqueous solutions can often be avoided by recourse to spectrometric determinations (Chapter 4) or, where absorbance characteristics are unfavourable, the indicator variant of spectrophotometry (p. 94), or the refinements of potentiometric technique which permit titrations down to 0·0002M (Chapter 3).

In the pharmaceutical industry, poorly soluble substances are much titrated in partly aqueous solvents. The aim is not to find the aqueous pK_a value, but to place a series of candidate drugs in a ranking order which, should it correspond to an increasing physiological response, will indicate what compounds should next be synthesized. In such a situation, allowance can often be made for interference occasioned by lipophilic or steric properties if these have already been encountered in a multiple regression analysis (Hansch, 1971). Working along these lines, Simon (1964) has amassed pK-like values for several hundred substances in a 4:1 methoxyethanol: water mixture. Cookson (1974), who prefers a 1:1 mixture of these solvents, has stated that a collection of these values has value for the drug designer.

3 Refinements of potentiometric titration: apparatus and calculations

A APPARATUS

3.1 Semi-micro titrations

The method described in Chapter 2 can be scaled down, with no loss of accuracy and with a small gain in speed. It can be used for as little as one twenty-thousandth of a mole dissolved in 10 ml water (0·005M). This effects a great economy of material. A balance weighing accurately to 10 μg (i.e. a good 'five-place' balance) is required, also a micrometer syringe to deliver 0·5 ml of 0·1N titrant in 0·050-ml portions. The minimal volume of solution that can be titrated is 10 ml (in a 30-ml beaker): no smaller volume can, usually, cover the standard electrode assembly shown in Fig. 2.1. We have found this method very useful for routine determinations on new substances and have used it regularly since 1954. For more concentrated solutions, or for smaller fractions of a mole, a micro titration should be performed.

Semi-micro titrations of increased accuracy can be effected if special attention is given to temperature control. To this end, a suitable titration vessel is shown in Fig. 3.1. It is a tall beaker sealed into a water jacket through which water

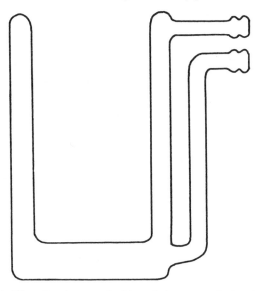

Figure 3.1 Glass titration vessel with sealed-on water jacket (in section).

from a thermostatically controlled bath is circulated so as to maintain the solution in a beaker at the required temperature $\pm 0.02°C$. The electrode assembly is as in Fig. 2.1 except that magnetic stirring is preferred to stirring by nitrogen bubbles. Stirring is continued for 1 min after each addition, but discontinued while readings are being taken. If an atmosphere of nitrogen is required, moist and oxygen-freed gas should be directed across the surface of the liquid.

To achieve a steady performance, the pH measuring unit and the constant temperature bath should be turned on at least 12 h earlier. The membrane of the glass electrode should have been conditioned by soaking it in water for about 14 days, and this electrode should be used in conjunction with a calomel reference half cell of the porous plug type. If the glass electrode has a 'Zero point pH of 7' (i.e. its bulb contains pH 7 buffer plus chloride) the electrodes can best be stored by dipping them into pH 6.8 phosphate buffer maintained at constant temperature in the bath. The level of the potassium chloride in the calomel half cell should be below that of the upper surface of the rubber bung by which the half cell is retained (using 'O'-ring arrangements, as mentioned on p. 19), but at least 2 cm above the level of the buffer solution in the beaker. The beaker should be held by a rack fitted into the bath so that the level of the bath water is about 2 cm below the rim of the beaker. Such an arrangement ensures that the potassium chloride solution of the half cell is maintained at the same temperature as the solution, and that a slight hydrostatic pressure differential exists between it and the buffer solution. The bung holding the electrodes should be bored similarly to the one described in Chapter 2, p. 19, and should be of such a diameter that it can be transferred easily between beakers mounted in the bath and the jacketed beaker in which the titration is performed. An extra rubber sleeve is sometimes useful in this regard, and we make use of the sleeves taken from a bung developed originally for glassblowers. This device consists of a series of conically shaped rubber sleeves which, when assembled, the one inside the other, constitute a bung.

The stability of the cell should be checked by dipping the electrodes into a fresh portion of the phosphate buffer solution maintained at constant temperature. When transferring the electrodes from one solution to another, each electrode is washed three times with the solution into which it will be placed. These washings are discarded and no attempt should be made to dry the electrodes. After allowing about 8 min for stabilization, the pH of the phosphate buffer is either monitored continually by chart recorder or is measured at frequent intervals for a minimum of 2 hours. The drift in pH should not exceed 0.002 pH hour^{-1}, and in our experience it usually exceeds this specification by being less than 0.0003 pH hour^{-1}, when measured by an independent potentiometer over a 24-hour period. This check need only be performed when a glass electrode or a calomel half cell is being used for the first time. However, a linearity check should be carried out at the beginning of each working day by transferring the electrodes several times between fresh portions of phosphate buffer solution and an appropriate pH reference solution selected from Table 2.1. After the initial standardization, the reproducibility of the pH values should be found to be closer than 0.005 pH unit.

Before transferring the electrodes to the water-jacketted titration beaker, they should be washed with water the temperature of which is the same as that of the water bath. No attempt should be made to dry the electrodes, but superficial water should be gently removed by means of a paper tissue.

During a determination, the titrant is added by a micrometer syringe to which is attached a length of polythene cannula tubing. This is immersed in the solution only during the delivery of the titrant and is withdrawn during the measurement of pH. This procedure is tedious, particularly when the total volume of titrant is large, say 10 ml. However this precaution does hasten the attainment of a steady reading, which is always slow after stirring has been halted. During this period, the syringe can be refilled. The time taken to reach an equilibrium pH reading varies with the concentration chosen for the determination, and the ionic strength. The maximum time we have observed was 16 min during the early stages of titration of a very dilute solution. To detect when this equilibrium is reached, we use a pen recorder attached to the pH meter for both standardization and measurements. Invariably a trend to higher pH values occurs after the cessation of stirring. The time taken to complete a titration is about 50 min when one equivalent of titrant is added in 0.05 equivalent portions to a solution whose concentration is 0.005M or greater. The titration of succinic acid, for the example recorded in Table 3.3 on p. 60, required 90 min to complete at 0.005M, and about 150 min at 0.0025M, or lower.

The accuracy of the method using these refinements is discussed on p. 63. After the titration, the electrodes must be returned promptly to the phosphate buffer, and the equilibrium value of pH measured. This value should be within ± 0.005 of the original standardized value. Before commencing another titration, the linearity of the electrodes should be checked again, and adjusted if necessary.

3.2 Micro titrations

Although commercial apparatus is available for micro titrations, much of this is fragile or hard to clean. The apparatus shown in Fig. 3.2 can be put together from easily accessible materials plus a miniature glass electrode. With its aid as little as 0.5 mg of a substance, dissolved in 0.5 ml water, can readily be titrated.

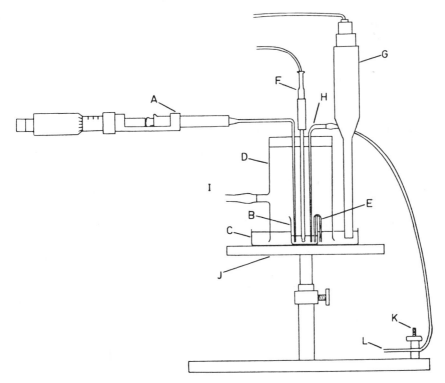

Figure 3.2 Apparatus for micro-determination of ionization constants by potentiometry. A, Micrometer syringe. B, Titration vessel. C, Petri dish containing saturated KCl solution. D, Glass shield and stopper (supported as described on p. 19). E, Salt bridge. F, Miniature glass electrode. G, Calomel electrode (full size). H, Nitrogen delivery tube for stirring. I, Nitrogen delivery tube for supplying an inert atmosphere. J, Adjustable platform. K, Precision pinchcock. L, Entry for moist, oxygen-free nitrogen from aspirator shown in Fig. 3.4.

In this way, ionization constants can be used to help identify small fractions isolated in chromatography, and the solution remaining after titration can be used for ultraviolet spectroscopy. For a substance of mol.wt. 200, the above concentration becomes 0·005M, and is suitable for obtaining a pK_a that is not below 2·3 (see Chapter 2, p. 25).

The apparatus of Fig. 3.2 is suitable for volumes of solution from 0·5 to 5·0 ml. It should be located in a thermostatically controlled room. A petri dish, 9 cm diameter and containing saturated potassium chloride solution, is placed centrally on the adjustable platform. The titration vessel is a flat-bottomed tube of glass or polythene. It should be 20–30 mm high and have a diameter in the range 10–20 mm according to the volume required. Electrical contact between the solution to be titrated and the potassium chloride solution in the dish is made by a salt bridge constructed as in Fig. 3.3. Thus, glass tubing of 7 mm diameter is drawn out to about 1 mm diameter. A U-shaped bend is made by carefully warming this capillary tube over a pilot flame so that one limb is about 30 mm longer than the other. The distance between the limbs should be 2 to 3 mm. A right-angled bend is made where the thin and thick portions of the tubing meet. The longer limb is placed in a warm solution of agar (0·3 g) dissolved in saturated aqueous potassium chloride (10 ml) and a slight suction is applied so that the solution fills the capillary completely. When the agar is cool and set, the limbs are cut to an equal length. The bridge is suspended on the rim of the titration vessel. Such bridges are stored in saturated potassium chloride solution when not in use. The level of the solution in the petri dish should be adjusted so that it is about 2 mm above the level of the solution in the titration vessel.

The miniature glass electrode is supported above the adjustable platform in a glass shield by a rubber bung from which a 10° sector has been removed to admit the micrometer syringe needle used in titration. A 3-mm diameter hole is drilled in the bung to take a glass capillary used to distribute nitrogen for stirring. Both the needle and the capillary dip into solution to be titrated. The glass shield is a cylinder, about 35 mm diameter and 55 mm high. It is fitted with a side arm through which nitrogen can be passed if an inert atmosphere is required because the amount of nitrogen used in stirring is not enough for this

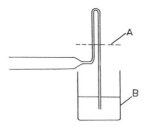

Figure 3.3 Filling the salt bridge. A, Cut here after filling. B, Solution of agar in aqueous potassium chloride.

purpose. The glass electrode is positioned so that the tip is 1 mm above the lower edge of the shield and about 50 mm above the top of the petri dish when the platform is in the fully lowered position. The shield and the electrode are rigidly held by clips (e.g. Terry clips) which are screwed into the ends of pieces of wooden dowelling which can be held, by bossheads, to a small retort stand (not shown in Fig. 3.2). If the apparatus is used frequently, a further refinement is to supply rack-and-pinion movements to control the height of the platform and to locate and lower the micrometer syringe.

The calomel electrode is attached in a similar way to the same vertical support. It is positioned outside the glass shield so that its leak is at the same level as the lower edge of the glass shield.

The substance whose ionization constant is to be determined is weighed directly into the tared titration vessel. Water is added and the substance is dissolved (by careful warming if necessary). After cooling, the volume of the solution is determined by weighing the vessel and its contents to the nearest milligram.

The miniature glass electrode is checked for linearity with phthalate and borate buffers as described on p. 17. The electrode is then rapidly washed.

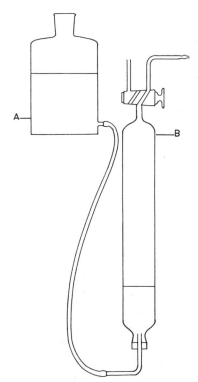

Figure 3.4 Aspirator to supply nitrogen to the micro-apparatus. A, water. B, nitrogen reservoir (500 ml).

This washing is conveniently done with an ungraduated pipette of about 15-ml capacity fitted with a rubber bulb. A 45° bend about 15 mm from the end of the capillary enables a jet of water to be directed upwards towards the glass electrode. The adhering droplets can be removed from the electrode and shield with small pieces of filter paper. The titration should be begun immediately after standardization and washing are complete. To this end, the salt bridge is placed in position and the vessel is raised (by raising the petri dish) so that the bulb of the glass electrode dips into the solution.

Purified nitrogen gas from a reservoir of about 500-mm capacity (Fig. 3.4) is passed through a narrow-bore polythene tube to the stirring capillary. The rate of flow is controlled at about 2 bubbles per second by a precision pinch-cock (e.g. the Pregl type). The flow of nitrogen is maintained whilst each portion of the titrant is being added and for about one minute longer. The flow is stopped, by closing the stopcock on the reservoir, before reading the pH. Equilibrium is reached in about one minute after the nitrogen ceases to flow.

If a thermostatically controlled room is not available, the petri dish should be jacketted (i.e. made double-walled). Water from a thermostatically controll-ed bath can be circulated through the jacket to maintain the temperature in the titration vessel as required $\pm 0.5°C$.

3.3 The rapid-flow method

When a substance is stable in one ionic species, but unstable in another, a rapid mixing of the solution with a stoichiometric amount of the titrant is not difficult to arrange shortly before the mixture streams past the electrodes. Such an arrangement has been described for amidines (Schwarzenbach and Lutz, 1940) which are stable as cations, but unstable as neutral molecules. Because of the high pH values involved, this example uses the hydrogen electrode; but at lower pH values, the glass electrode can be used instead.

3.4 The hydrogen electrode

This consists of a wire or plate, made of gold or platinum, on which has been deposited a very thin, catalytically active layer of platinum. These electrodes may be bought, or else made according to Hildebrand (1913). They require special care in maintenance (Kolthoff and Laitinen, 1941). Highly accurate determinations of pK_a can be made with the hydrogen electrode, in cells without liquid junction (Harned and Ehlers, 1932; Harned and Owen, 1958).

In general, the hydrogen electrode is cumbersome, and less useful than the glass electrode because it is easily poisoned by compounds of sulphur, mercury or arsenic. Also it may hydrogenate the substance being investigated. It is most used for potentiometric titration of substances having a pK_a in excess of 11, i.e. in a region where the glass electrode is likely to be inaccurate. However, it is useless in the presence of oxidizing or reducing substances for which it would record the oxidation–reduction potential.

The apparatus is set up as in Fig. 2.1 (p. 20), but with the hydrogen electrode replacing the glass electrode, and a stream of hydrogen (replacing the stream of nitrogen) directed at the tip of the hydrogen electrode. (In another type of electrode the hydrogen enters the glass sheathing of the platinum wire.) The stopper of the titration vessel should be tight-fitting, and an extra opening should be provided for a tube that will convey the excess gas, through a trap, to the open air outside the building.

For use above pH 11, every trace of oxygen must be removed from the hydrogen, by passing it either over platinized asbestos at 500°C, or through a (commercial) chromous salt apparatus which works at room temperature.

The titration vessel should be placed in a thermostat, and the cell checked against potassium hydrogen phthalate buffer. Although nothing so elaborate as a commercial pH set is required to obtain readings from this electrode assembly, such a set is always present in the laboratory and most models can conveniently be used as a millivolt potentiometer for the purpose.

For an example of the use of the hydrogen electrode in the pK_a 12–14 range see Thamsen (1952). Because of the high sensitivity of the hydrogen electrode to temperature variations, it is imperative to standardize the electrodes at the special temperature at which the titration is conducted.

Micro-titrations with the hydrogen electrodes can be made in a small, horizontally elongated cell of capacity 0·15 ml (Ogston and Peters, 1936) with three tubulatures. Hydrogen enters by one tubulature, flows over the surface of the solution, passes the centrally situated electrode and escapes through the last tubulature to the open air, via a trap. Contact of the electrode with hydrogen, as well as adequate mixing, is assured by keeping the cell mechanically rocked through an angle of 20° about three times a second. The calomel reference electrode is connected through a capillary and stopcock attached to the bottom of the cell. The apparatus is maintained in an air thermostat at the required temperature $\pm 0\cdot2$°C.

B CALCULATIONS

3.5 Monofunctional acids and bases

A small discrepancy exists between the pK_a values calculated in the tables of Chapter 2 and the standard values in the literature. For example, the pK_a value for benzoic acid, calculated in Table 2·4 is 4·16 $\pm$ 0·01 at 20°C, (similarly we obtained 4·16 $\pm$ 0·01 at 25°C), whereas most highly accurate values lie around 4·20 at 25°C (Kortüm *et al.*, 1961; Serjeant and Dempsey, 1979). The reason for this discrepancy is that our calculations, up to this point, have included *both* concentration and activity terms, and our results for benzoic acid are 'mixed pK_a' values. These mixed constants, which we call K_a^M, are constant only for the concentration of ionized species at which the pH was measured and stand in contrast to equation (3.1) in which all the terms are expressed as activities. The standard values for benzoic acid, on the other hand, have been calculated using activities throughout and the thermodynamic constant, K_a^T, is entirely independent of concentration (see p. 5 and Appendix II).

$$K_a^T = \frac{\{H^+\}\{A^-\}}{\{HA\}} \tag{3.1}$$

We shall now describe how activity corrections can be applied to data obtained from the potentiometric titration of monobasic acids and monoacidic bases, where the principal term in need of correction is the concentration of ionized species. This is because pH, as measured potentiometrically, is an activity rather than a concentration term (cf. Table 11.4), and the activity of the non-ionized species, $\{HA\}$ or $\{B\}$, is approximately equal to the concentration. The conversion of the concentration of the ionized species into its activity is based on the equations

$$\{A^-\} = [A^-]y_{A^-} \text{ for acids} \tag{3.2}$$

and

$$\{BH^+\} = [BH^+]y_{BH^+} \text{ for bases,} \tag{3.2a}$$

where y_{A^-} and y_{BH^+}, known as the activity coefficients of the ionized species, are usually less than one. In some circumstances (listed on p. 50) it is desirable to convert K_a^M to K_a^T. A few words on activity theory will now follow so that the nature of the correction process can be understood.

The general formulation of the relationship between the concentration of of an ion C_i and its activity coefficient y_i is based on the work of Debye and Hückel (1923, 1924). An important term in this formulation is the *ionic strength*, written as I and defined thus:

$$I = 0.5\Sigma C_i z^2 \tag{3.3}$$

where C_i is the molecular concentration of an ion and z is its valency and the symbol Σ denotes summation. According to Debye and Hückel, the activity coefficient y_i of an ion of valence z is related to the ionic strength by

$$-\log y_i = \frac{A z^2 I^{1/2}}{1 + B a_i I^{1/2}} \tag{3.4}$$

The terms A and B are constants which vary with the dielectric constant and temperature of the solvent. The values* of A are 0.5070 and 0.5115 $\text{mol}^{-1/2}$ $\text{litre}^{1/2}$ at 20° and 25°C, respectively, and $B = 0.3282$ and 0.3291×10^8 cm^{-1} $\text{mol}^{1/2}$ $\text{litre}^{1/2}$ at the corresponding temperatures. The term a_i is the ionic size parameter, i.e. the mean distance of approach of the ions, for which

*See Robinson and Stokes (1959) for a more extensive range of values.

4.5×10^{-8} cm may be taken as an average value. At $25°$ and $20°C$, equation (3.4) becomes

$$- \log y_i = \frac{0.512 I^{1/2}}{1 + 1.5 I^{1/2}} \quad \text{and} \quad \frac{0.507 I^{1/2}}{1 + 1.5 I^{1/2}},$$

respectively. The ionic strength term in these equations needs further definition. When a monobasic acid is dissolved in water the only ions in the solution are H^+, A^-, and (to a lesser degree) OH^-. When this solution is titrated with potassium hydroxide, the ionic strength at any point during the titration is

$$I = 0.5([K^+] + [H^+] + [A^-] + [OH^-]).$$

If, for $[A^-]$, the term $[H^+] + [K^+] - [OH^-]$, as derived in equation (2.4a), is substituted,

$$I = [K^+] + [H^+] \tag{3.5}$$

where $[K^+]$ equals the concentration of potassium hydroxide added. Similarly, for a titration of a conjugate base (i.e. an anion) with hydrochloric acid (see Table 2.5), the ionic strength is also given by equation (3.5), but $[K^+]$, or in this case $[Na^+]$, is now the total concentration of the salt originally taken. The equation to be used for the titration of a base with acid or a back-titration of a cation with alkali is:

$$I = [Cl^-] + [OH^-] \tag{3.5a}$$

where $[Cl^-]$ is the concentration of hydrochloric acid added or, if a back-titration has been performed, the total concentration of the cation taken.

The thermodynamic ionization constant, defined by equation (3.1) can, therefore, be written as

$$K_a^T = \frac{\{H^+\}[A^-]}{\{HA\}} \cdot y_{A^-} \quad \text{for acids} \tag{3.6a}$$

and

$$K_a^T = \frac{\{H^+\}\{B\}}{[BH^+]} \cdot \frac{1}{y_{BH^+}} \quad \text{for bases,} \tag{3.6b}$$

in which the terms y_{A^-} and y_{BH^+} are the respective activity coefficients calculated by means of equation (3.4). Hence at $20°C$

$$pK_a^T = pH - \log \frac{[\text{non-protonated}]}{[\text{protonated}]} \pm \frac{0.507 I^{1/2}}{1 + 1.5 I^{1/2}}. \tag{3.7}$$

The last term is positive for acids and negative for bases. It follows from equation (3.7) that the ionization constants of mixed type (K_a^M), as obtained in Chapter 2, can be converted to thermodynamic ionization constants by the equation

$$pK_a^T = pK_a^M \pm \frac{0.507\,I^{1/2}}{1 + 1.5\,I^{1/2}},$$ (3.7a)

the sign of the activity correction term being allocated as in equation (3.7). Thus a trend is expected in the values of K_a^M (as given in the tables of Chapter 2) when the ionic strength of the solution is increased by increasing neutralization. For example, a monoacidic base, titrated at a concentration of 0.01M, requires a correction subtracted from pK_a^M of 0.015 when $[BH^+]$ is 0.001M, and 0.041 when $[BH^+]$ is 0.009 (e.g. 'Tris' in Table 2.7). However, this trend is negligible if, as in routine pK determinations, the measurements are performed with apparatus calibrated in 0.02 pH unit. For this reason we were able to state, in Chapter 2, that for many purposes titrations may be carried out without activity corrections provided that the concentration does not exceed 0.01M. It is important to note that in these examples the concentrations of hydroxyl or hydrogen ions were small in comparison with the concentrations of ionized species. In such examples where the concentrations of hydrogen or hydroxyl ions are negligible and the actual concentrations of HA and A^- are equal to their stoichiometric concentrations, the thermodynamic pK values can be obtained approximately by setting a_i equal to zero and considering the ion as a point charge. Thus, for solutions more dilute than 0.01M, equation (3.7a) can be approximated to

$$pK_a^T = pK_a^M \pm 0.5\,I_m^{1/2}$$ (3.8)

where I_m is the ionic strength at the mid-point of the titration (i.e. for a 0.01M titration, I_m is 0.005). If a back titration has been performed, the ionic strength can be taken as remaining at the total concentration of substance.

The question must now be asked: When is it advisable to use activity corrections in calculating the results from a potentiometric titration? Principally when work of the highest precision is being undertaken. Hence we recommend that these corrections be applied to all results when an instrument calibrated in 0.005 unit or less has been used and the refinements of technique given on p. 41 have been employed. When using apparatus which can be read only to 0.02 or 0.01 pH units, it may be profitable to apply activity corrections only in the following three cases:

(a) When the concentration lies between 0.005 and 0.1M*;
(b) When the hydrogen ion or hydroxyl ion concentration is comparable in magnitude to the stoichiometric concentration of the ionized species (see p. 27); or
(c) When the weaker acidic group of a dibasic acid, or the weaker basic group of a diacidic base, is titrated at a concentration greater than 0.0025M (see p. 53).

*0.1M is commonly considered an upper limit beyond which activity corrections, as commonly applied, are not valid.

3.6 Method of calculation

This must allow for the dilution of the initial concentration of substance caused by the addition of titrant. For this purpose the concentration, C, is given by

$$C = \frac{W \cdot 10^3}{M(V + V_T)}, \tag{3.9}$$

where W is the weight taken, M is the molecular weight, V is the volume of water used for dissolving the substance, and V_T is the volume of titrant added. The concentration of K^+ or Cl^- in the solution must be calculated similarly by

$$C_A = \frac{V_T \cdot N}{(V + V_T)} \tag{3.9a}$$

where C_A is the concentration of K^+ or Cl^- and N is the normality of the acid or alkali added.

A difficulty arises in the calculation of the ionic strength if the substance is a moderately strong monobasic acid ($pK < 4$), or a moderately strong mono-acidic base ($pK > 10$). It should be noted that equations (3.5) and (3.6) require the use of *concentration* terms, for the hydrogen ion in (3.5) and the hydroxyl ion in (3.6). However, the only terms initially available for substitution in these equations will be the *activities* of these ions calculated from the measured hydrogen ion activity (i.e. pH). The ionic strength must, therefore, be approximated initially to $I = [K^+] + \{H^+\}$ for acids and $I = [Cl^-] + \{OH^-\}$ for bases in order to calculate the activity coefficient by equation (3.4). These approximate activity coefficients are then used to calculate the hydrogen ion or hydroxyl ion concentration (i.e. $[H^+] = \{H^+\}/y$; or $[OH^-] = \{OH^-\}/y$) by equation (3.4); and also the ratios of the activities of the non-protonated to protonated species. For acids the ratio is

$$\frac{\{A^-\}}{\{HA\}} = \frac{(C_A + [H^+] - [OH^-]) \cdot y_{A^-}}{(C - C_A + [OH^-] - [H^+])}, \tag{3.10}$$

and for bases

$$\frac{\{B\}}{\{BH^+\}} = \frac{(C - C_A - [OH^-] + [H^+])}{(C_A + [OH^-] - [H^+]) \cdot y_{BH^+}}. \tag{3.10a}$$

For *weak acids* and *weak bases* these ratios can be used directly to calculate the pK_a value because the 'approximate value' of the ionic strength is, in these cases, equal to its true value and hence the activity coefficients will also be correct. This is not the case for the *stronger acids and bases* for which values of the activity coefficients should be recalculated using the values of $[H^+]$ or $[OH^-]$ in the calculation of the ionic strength. This process should be repeated until successive values of the ratios calculated from equations (3.10) and (3.10a) are constant. If this method of successive approximations has to be employed, the calculations take a great deal of time.

Fortunately there is an empirically derived approximation, suitable for results obtained with a set calibrated in 0·02 pH unit, e.g. for bases:

$$pK_a^T = pH + \log \frac{[BH^+] \cdot y_i - \{H^+\}}{[B] + \{H^+\}} \tag{3.10b}$$

where y_i can be obtained from equation (3.4). Thus very little calculation is needed beyond that used in Table 2.8. Two extra columns should be added to those used in that table. The first of these columns should give the total concentration of base, allowing for volume changes caused by the addition of the titrant; the second column should list values of $[BH^+] \cdot y_i$.

The following data permit comparison of pK_a^T values calculated, alternatively, by equations (3.10a) and (3.10b), at two different concentrations (each result is the mean of nine values):

Conc. (M) (of purine)	pK_a^T from (3.10a)	pK_a^T from (3.10b)
0·05	2·43 ± 0·01	2·44 ± 0·02
0·02	2·44 ± 0·03	2·44 ± 0·02

Let us consider a still weaker univalent base, one of pK_a^T 1·30, titrated at a concentration of 0.05M. The error caused by using the approximation (3.10b)

Table 3.1   Comparison of the experimentally determined constants (pK_a^M) and the thermodynamic constant (pK_a^T) of β-naphthol at 20°C

1	2	3	4	5
Concentration (M)	Ionic strength (I_m)	pK_a^M	pK_a^T	Δ
0·022 14	0·011 07	9·583	9·632	0·049
0·041 64	0·020 82	9·566	9·631	0·065
0·061 00	0·030 50	9·555	9·632	0·077
0·080 22	0·040 11	9·545	9·631	0·086
0·099 28	0·049 64	9·539	9·633	0·094
0·146 28	0·073 14	9·521	9·631	0·110
0·192 40	0·096 20	9·511	9·631	0·120
			Result: 9·632 ± 0·001	

is 0·03 with reference to the pK_a^T as calculated by equation (3.10a). This error is no greater than that produced by an error of 0·01 pH in the reading from the pH set. We have had much experience with equation (3.10b) and we recommend it.

Results

The results in column 3 of Table 3.1 show a decline in the pK_a^M value of β-naphthol from 9·583 to 9·511 with a rise of ionic strength from 0·011 to 0·096. The pK_a^T in column 4 was obtained by extrapolation of the straight line produced by plotting $I^{1/2}/(1 + I^{1/2})$ against pK_a^M. A similar discussion for acetic acid is given on p. 196.

3.7 Diacidic bases, dibasic acids and ampholytes

Substances of these types have two ionizing groups. For diacidic bases and dibasic acids both groups can be titrated sequentially with two equivalents of titrant. The results from the first equivalent can be calculated in exactly the same way as was given in this Chapter for monoacidic bases and monobasic acids. For the second ionization process, these calculations are not applicable because the ionic strength term is different and the activity coefficients much smaller than for the first ionization process. This is predictable from equations (3.3) and (3.4) in which there appear terms for the square of the valency (i.e. z_i^2) and for the ionizations

$$H_2B^{2+} \rightleftharpoons BH^+ + H^+ \text{ for bases,} \qquad (3.11)$$

and

$$HA^- \rightleftharpoons A^{2-} + H^+ \text{ for acids,} \qquad (3.11a)$$

the species H_2B^{2+} and A^{2-} are both divalent. *Ampholytes*, on the other hand, give rise to singly charged ions, provided that their ionization processes are well-separated; hence two solutions should be prepared for the determination of the pK_a values. The first is titrated with acid to yield the basic pK_a of the basic group and the second with alkali to yield the pK_a of the acidic group. These pK_a values are calculated just as they would be for separate monofunctional acids and bases.

We will now discuss the determination of the pK_a values for the ionization processes described by equations (3.11) and (3.11a) by deriving the equations for a diacidic base. The corresponding equations for a dibasic acid will then be stated.

The charge balance when the weaker group of a diacidic base is ionizing can be expressed as

$$2[H_2B^{2+}] + [BH^+] + [H^+] = [Cl^-] + [OH^-]$$

and the total concentration, C, of the base as

$$C = [H_2B^{2+}] + [BH^+].$$

Thus

$$[H_2B^{2+}] = [Cl^-] + [OH^-] - [H^+] - C \qquad (3.12)$$

and hence

$$[BH^+] = 2C - [Cl^-] - [OH^-] + [H^+]. \qquad (3.12a)$$

The use of equation (3.3), to define the ionic strength, gives

$$I = 0.5(4[H_2B^{2+}] + [BH^+] + [H^+] + [OH^-] + [Cl^-]). \qquad (3.13)$$

Substituting the values of $[H_2B^{2+}]$ and $[BH^+]$, given by equations (3.12) and (3.12a) in equation (3.13),

$$I = 2[Cl^-] + 2[OH^-] - [H^+] - C. \qquad (3.14)$$

The thermodynamic ionization constant for the process described by equation (3.11) is

$$K_a^T = \frac{\{H^+\}[BH^+]}{[H_2B^{2+}]} \cdot \frac{y_{BH^+}}{y_{H_2B^{2+}}}. \qquad (3.15)$$

The activity coefficients can be expressed in exponential form by defining a function, FS, of the ionic strength as

$$FS = \frac{I^{1/2}}{1 + 1.5I^{1/2}}, \qquad (3.16)$$

whereupon

$$y_i = \frac{1}{10^{(Az^2FS)}}.$$

At 25°C therefore,

$$y_{BH^+} = \frac{1}{10^{(0.5115FS)}} \quad \text{and} \quad y_{H_2B^{2+}} = \frac{1}{10^{(2.046FS)}}. \qquad (3.17)$$

It follows that equation (3.15) can be written as

$$K_a^T = \frac{\{H^+\} \cdot [BH^+]}{[H_2B^{2+}]} \cdot 10^{(1.5345FS)}.$$

Substituting the values for $[BH^+]$ and $[H_2B^{2+}]$ yields the equation

$$pK_a^T = pH + \log\frac{([Cl^-] - C + [OH^-] - [H^+])}{(2C - [Cl^-] - [OH^-] + [H^+])} - \frac{1.5345\,I^{1/2}}{1 + 1.5I^{1/2}}. \qquad (3.18)$$

The term $[Cl^-]$ is the total concentration of hydrochloric acid added *including* that used in the neutralization of the stronger group. This applies also to equation (3.14) for the calculation of the ionic strength. In both the equations, $[H^+]$ and $[OH^-]$ should be expressed as concentrations. For this purpose

$[H^+] = \{H^+\} \cdot 10^{0.5115FS}$ where $\{H^+\}$ is calculated from the measured pH.
A similar derivation for the weaker group of a dibasic acid yields

$$I = 2[K^+] - C + 2[H^+] - [OH^-] \tag{3.19}$$

and, at 25°C,

$$pK_a^T = pH + \log\frac{(2C - [K^+] - [H^+] + [OH^-])}{([K^+] - C + [H^+] - [OH^-])} + \frac{1.5345\,I^{1/2}}{1 + 1.5\,I^{1/2}}. \tag{3.20}$$

In this case $[K^+]$ is the total concentration of potassium hydroxide added, *including* that used in the neutralization of the stronger group. The terms $[H^+]$ and $[OH^-]$ are calculated as described above.

An equation analogous to (3.7a) for the ionizations of the weaker groups of diacidic bases or dibasic acids is

$$pK_a^T = pK_a^M \pm \frac{1.5345\,I^{1/2}}{1 + 1.5\,I^{1/2}} \tag{3.21}$$

at 25°C. The last term is positive for acids and negative for bases.

The results obtained by titrating the weaker basic group of ethylenediamine are shown in Table 3.2, and test the validity of equation (3.21). This was rearranged for this purpose into an equation* of the form $y = mx + c$

i.e. $$pK_a^M = pK_a^T + \frac{m \cdot I^{1/2}}{1 + 1.5\,I^{1/2}} \tag{3.21a}$$

where m is the slope (theoretically 1.521 at 20°). Instead of solving this equation

Table 3.2 Comparison of mixed and thermodynamic ionization constants for the titration of the second step of a diacidic base

Example: ethylenediamine (20°C)

I	x values = $I^{1/2}(1 + 1.5I^{1/2})^\S$	y values = pK_a^m	x^2	$x \cdot y$	pK_a^T	Δ
0.005 00	0.0639	7.136	0.4083×10^{-2}	0.455 99	7.044	0.092
0.014 86	0.1031	7.192	1.0630	0.741 50	7.043	0.149
0.024 50	0.1268	7.222	1.6078	0.915 75	7.039	0.183
0.043 26	0.1585	7.272	2.5122	1.152 61	7.043	0.229
0.061 32	0.1806	7.304	3.2616	1.319 10	7.043	0.261
0.078 70	0.1974	7.328	3.8967	1.446 55	7.043	0.285
0.095 46	0.2111	7.348	4.4563	1.551 16	7.043	0.305
Totals	$\Sigma x = 1.0414$	$\Sigma y = 50.802$	$\Sigma x^2 =$	$\Sigma xy =$		
	$(\Sigma x)^2 = 1.0845$		17.206×10^{-2}	7.582 66		

Results: $pK_a^T = 7.042 \pm 0.003$ at 20°C

§ $- 3 \log y_i = 1.52 \times$ this term.
*This use of y is algebraic, and does not refer to the activity coefficient.

graphically by plotting values of pK_a^M (y-axis) against the corresponding values of $I^{1/2}/(1 + 1 \cdot 5 I^{1/2})$ (x-axis), the *method of least squares* was applied. This algebraic method, which is used elsewhere in this book, will now be described. The slope, m, of the linear relation is given by

$$m = \frac{n\Sigma(x \cdot y) - \Sigma x \cdot \Sigma y}{n\Sigma(x)^2 - (\Sigma x)^2}$$

and the intercept, c, by

$$c = \frac{\Sigma(x)^2 \cdot \Sigma y - \Sigma x \cdot \Sigma(x \cdot y)}{n\Sigma(x)^2 - (\Sigma x)^2},$$

where n is the number of observations. In this case we assign the values of pK_a^M to y and the values of $I^{1/2}/(1 + 1 \cdot 5 I^{1/2})$ to x and (for this example) n in the above equations is equal to 7. The method calls for the following steps which can be followed by reference to Table 3.2. First, the individual values of $I^{1/2}/(1 + 1 \cdot 5 I^{1/2})$ are added to yield the sum, Σx. The squares of these individual values are summed also to obtain Σx^2. Next, the observed values of pK_a^M for the given values of the ionic strength are summed, forming the total Σy. These individual values are also multiplied by the corresponding values of $I^{1/2}/(1 + 1 \cdot 5 I^{1/2})$ and the individual products summed to yield $\Sigma x \cdot y$. These totals are inserted in the above equations to yield the intercept $pK_a^T = 7 \cdot 042$, and the slope $1 \cdot 45$. The latter should be equal to $3 \times 0 \cdot 507$ (see p. 48) and the small discrepancy is due to the uncertain value of a_i in equation (3.4).

In calculating ionic strength from equations (3.14) (for bases) and (3.19) (for acids), the weaker group of a base or an acid is seldom found to be sufficiently strong to make the respective terms $2[OH^-]$ and $2[H^+]$ significant. Corrections for the other terms, $[H^+]$ for bases and $[OH^-]$ for acids, should be made by the method of successive approximations as described on p. 51. It must also be borne in mind that activity corrections, applied as described in this chapter, are not valid if the ionic strength exceeds $0 \cdot 1M$. For this reason we recommend that the total concentration of solutions of bifunctional compounds should not exceed $0.025M$.

3.8 Overlapping ionization processes

If, in titrations of these classes of compounds, it is observed that the pK_a values (calculated by the methods already described) are of poor precision, it is very probable that the endpoint of the first equivalent is indefinite because the titration of one group has begun before that of the other is complete. When this occurs, the two ionization processes are said to overlap and the two pK_a values will be found to be separated by less than $2 \cdot 7$ units of pK_a. Speakman (1940) derived an equation which allows these pK values to be calculated.

It can be written in the form

$$\frac{1}{K_1^M} \cdot \frac{\{H^+\}^2 \cdot F}{(2-F)} - K_2^M = \{H^+\} \cdot \frac{1-F}{2-F} \tag{3.22}$$

where

$$F = \frac{C_A + \{H^+\} - \{OH^-\}}{C}. \tag{3.22a}$$

The terms C_A and C are defined by equations (3.9a) and (3.9) respectively. The hydrogen and hydroxyl ion activity terms have been placed in the equation to emphasize that these are assumed to be the measured quantities derived from the pH measurements. Whilst it is correct to use the hydrogen ion *activity* in equation (3.22), we shall need to replace it by hydrogen ion *concentration* in equation (3.22a). It is apparent, therefore that mixed constants would result from the solution of equation (3.22). A further complication is the ionic strength which, in this case, cannot be calculated directly from the stoichiometric concentration used, for example, in equation (3.19). To obtain the thermodynamic values, K_1^T and K_2^T, the activity functions must be calculated from estimates of the ionic strength. How this can be achieved will now be described for dibasic acids followed by a similar method for ampholytes and diacidic bases.

The ionization processes for a dibasic acid can be represented as

$$H_2A \overset{K_1}{\rightleftharpoons} H^+ + HA^-$$

$$HA^- \overset{K_2}{\rightleftharpoons} H^+ + A^{2-}$$

for which we can write

$$K_1^T = K_1^M \cdot y_{HA^-} \quad \text{and} \quad K_2^T = \frac{K_2^M \cdot y_{A^{2-}}}{y_{HA^-}}.$$

By defining a term FS (described by equation 3.16) and evaluating the activity coefficients, these equations become at 25°C

$$K_1^T = \frac{K_1^M}{10^{(0.5115 \cdot FS)}} \quad \text{and} \quad K_2^T = \frac{K_2^M}{10^{(1.5345 \cdot FS)}}.$$

It follows that the terms in equation (3.22) can be transformed to the corresponding thermodynamic constants to yield

$$\frac{1}{K_1^T} \cdot \frac{\{H^+\}^2 \cdot F}{2-F} \cdot \frac{1}{10^{(0.5115 \cdot FS)}} - K_2^T \cdot 10^{(1.5345 \cdot FS)} = \{H^+\} \frac{1-F}{2-F}.$$

Defining the terms in $\{H^+\}$ and F on the left-hand side as X and those on the right-hand side as Y, this equation can be written as

$$\frac{1}{K_1^T} \cdot \frac{X}{10^{(2.046 \cdot FS)}} - K_2^T = \frac{Y}{10^{(1.5345 \cdot FS)}} \tag{3.23}$$

This equation, in conjunction with equation (3.22), can be used to calculate the thermodynamic constants, K_1^T and K_2^T. The steps in the calculation are as follows:

1. Solve equation (3.22) by the method of least squares to obtain approximate values of K_1^M and K_2^M;
2. Calculate the concentrations of ionized species using these values of K_1^M and K_2^M in the equations

$$[HA^-] = \frac{K_1^M \{H^+\}}{G} \cdot C,$$

and

$$[A^{2-}] = \frac{K_1^M \cdot K_2^M}{G} \cdot C \tag{3.24}$$

where

$$G = \{H^+\}^2 + K_1^M \{H^+\} + K_1^M K_2^M;$$

3. The values $[HA^-]$ and $[A^{2-}]$ are then used to calculate the approximate ionic strength, which is given by

$$I = 0 \cdot 5(C_A + [HA^-] + 4[A^{2-}] + \{H^+\} + \{OH^-\}), \tag{3.25}$$

and hence the activity functions;
4. Convert the hydrogen ion activity to the corresponding concentration

$$[H^+] = \{H^+\} \cdot 10^{(0 \cdot 5115 \cdot FS)}$$

and use this term and the term $[OH^-]$ calculated from it in equation (3.22a) to recalculate F. Calculate also the activity corrections to be applied in equation (3.23);
5. Solve both equations (3.22) and (3.23) by least squares and use K_1^M and K_2^M from the former to calculate $[HA^-]$ and $[A^{2-}]$ as before. Substitute the latter term in equation (3.25) together with $[H^+]$, calculated above, to refine the values of the ionic strength;
6. Repeat the sequence of steps 4 and 5 until successive values of K_1^T and K_2^T are constant.

The thermodynamic ionization constants for ampholytes and diacidic bases can be calculated by a similar method. For these determinations, ampholytes are dissolved in one equivalent of hydrochloric acid and bases in two equivalents. The solutions are then back-titrated with two equivalents of potassium hydroxide.

For *ampholytes*, equation (3.23) becomes

$$\frac{1}{K_1^T} \cdot X - K_2^T = \frac{Y}{10^{(0 \cdot 5115 \cdot FS)}}$$

and for *diacidic bases* it is

$$\frac{1}{K_1^T} \cdot X \cdot 10^{(2 \cdot 046 \cdot FS)} - K_2^T = Y \cdot 10^{(0 \cdot 5115 \cdot FS)}.$$

The corresponding equations for the ionic strength, using the same nomenclature as equation (3.25) are as follows. For ampholytes, the equation is

$$I = 0 \cdot 5([Cl^-] + C_A + [H_2A] + [A^{2-}] + [H^+])$$

where

$$[H_2A] = \frac{\{H^+\}^2 \cdot C_t}{G}$$

and $[Cl^-]$ is equal to the concentration of hydrochloric acid in the solution before the start of the back-titration. G is defined in equation (3.24) and so is $[A^{2-}]$ which, for ampholytes, is the concentration of nonprotonated species. For diacidic bases, the equation is

$$I = 0 \cdot 5([Cl^-] + C_A + 4[H_2A] + [HA^-] + [H^+]).$$

$[H_2A]$ in this case is the concentration of diprotonated species (H_2B^{2+}) and $[HA^-]$ the concentration of monoprotonated species (BH^+). These terms are calculated in the manner described above. The time taken to perform these calculations is disproportionally greater than the time required to undertake the actual measurement. Fortunately, they may be adapted to a computer program, Table 3.5 at the end of this chapter, which conveniently illustrates the steps in the calculation. The time taken to process twelve sets of titration data was about 3 minutes and the program, written in FORTRAN IV, can be applied to data obtained at 20°C or 25°C. For help with computer programming, see Dickson (1968).

The results obtained from the titration of succinic acid (0.005M) are given in Table 3.3. The refinements of technique described on p. 42 were employed. The values of pK_1 and pK_2 obtained by Bates and Gary (1961), using a hydrogen electrode in a cell without liquid junction, were $pK_1 = 4 \cdot 209$, $pK_2 = 5 \cdot 638$ at 25·0°C.

Approximate methods
Those who have found the above approaches too time-consuming when they have needed only a reasonably close answer have usually titrated a monoester (when a dibasic acid has been their problem) to obtain the pK_a of the unesterified group. The pK_a of the other group was then obtained by noting the pH when one equivalent, only, of alkali had been added to the dicarboxylic acid, doubling this figure, and subtracting from it the pK_a already found. Some criticism of taking $-CO_2Me$ as the electronic equivalent of $-CO_2H$ will be found in Chapter 8. Neglect of activities when doubly charged ions are present, as here, also militates against a fine result.

A more rapid method is that of Martin (1971) who treated the unknown, in

Table 3.3   Determination of both ionization constants of a dibasic acid, requiring separation of overlapping pK_a values

Substance: Succinic acid $C_4H_6O_4 = 118 \cdot 1$
Concentration: 0·05910 g. Dissolved in 100 ml ion-free water
Temperature = 25·0°C.

Titrant 0·1N-KOH (ml)	pH	X	Y	I	pK_1	pK_2
1·000	3·677	0·519 73E−08	0·797 91E−04	0·144 58E−02	4·199	
1·250	3·767	0·408 31E−08	0·623 34E−04	0·161 70E−02	4·200	
1·500	3·853	0·321 84E−08	0·489 13E−04	0·180 89E−02	4·202	
1·750	3·932	0·258 87E−08	0·388 08E−04	0·201 32E−02	4·201	
2·000	4·009	0·207 80E−08	0·307 65E−04	0·223 08E−02	4·202	
2·250	4·081	0·169 23E−08	0·245 17E−04	0·245 48E−02	4·200	
2·500	4·153	0·136 74E−08	0·193 93E−04	0·269 02E−02	4·201	
2·750	4·223	0·110 82E−08	0·152 70E−04	0·293 27E−02	4·200	
3·000	4·291	0·901 94E−09	0·119 43E−04	0·318 07E−02	4·199	
3·250	4·361	0·724 02E−09	0·916 24E−05	0·344 17E−02	4·200	
3·500	4·431	0·579 17E−09	0·689 29E−05	0·371 18E−02	4·201	
3·750	4·498	0·468 59E−09	0·508 17E−05	0·398 36E−02	4·198	
4·000	4·569	0·371 22E−09	0·356 63E−05	0·427 44E−02	4·199	
6·000	5·135	0·569 79E−10	−0·141 75E−05	0·700 16E−02		5·634
6·250	5·204	0·456 69E−10	−0·159 90E−05	0·738 71E−02		5·634
6·500	5·273	0·367 14E−10	−0·174 06E−05	0·778 03E−02		5·634
6·750	5·342	0·296 28E−10	−0·185 26E−05	0·817 88E−02		5·634
7·000	5·412	0·239 11E−10	−0·193 91E−05	0·858 43E−02		5·635
7·250	5·480	0·195 96E−10	−0·202 21E−05	0·898 02E−02		5·632
7·500	5·554	0·157 29E−10	−0·207 09E−05	0·939 83E−02		5·634
7·750	5·629	0·126 84E−10	−0·211 65E−05	0·981 21E−02		5·635
8·000	5·708	0·101 57E−10	−0·215 23E−05	0·102 29E−01		5·636
8·250	5·789	0·818 05E−11	−0·219 83E−05	0·106 36E−01		5·633
8·500	5·881	0·638 62E−11	−0·222 09E−05	0·110 59E−01		5·634
8·750	5·981	0·493 81E−11	−0·225 40E−05	0·114 75E−01		5·632
9·000	6·099	0·365 61E−11	−0·227 48E−05	0·119 00E−01		5·632

Average: $pK_1 = 4 \cdot 200$
$\qquad$ $pK_2 = 5 \cdot 634.$

turn, with 0.5, 1.0, and 1.5 equivalents of acid, then took K_1 to be:

$$\{H^+\}_{0.5} - \{H^+\}_{1.5},$$

and then obtained K_2 in the same way as the ester-titrators (above). Results, described as reasonable though rough, were explored for the region pK_a 6–10.

3.9 Polyelectrolytes

In our experience, more than two overlapping pK_a values are rarely encountered among simple organic substances. However, much overlap is common in oligomers, including biopolymers. Hence we shall describe briefly how the potentiometric method can be applied to determine the ionization constants of acids having three or more overlapping pK_a values. The experimental technique is an extension of that already described. For example, a quadribasic acid should be titrated with four equivalents of alkali in 0·1 equivalent portions. The following equation, analogous to (3·22), is used to calculate the results.

$$K_1 \frac{\{H^+\}^3(1 - F)}{(4 - F)} + K_1 K_2 \frac{\{H^+\}^2(2 - F)}{(4 - F)} + K_1 K_2 K_3 \frac{\{H^+\}(3 - F)}{(4 - F)}$$

$$+ K_1 K_2 K_3 K_4 = \frac{\{H^+\}^4 \cdot F}{(4 - F)}, \tag{3.26}$$

where F is defined by equation (3.22a). This equation yields mixed constants (pK^M) and can be written in the form

$$ax + by + cz + d = w$$

where $a = K_1$, $b = K_1 K_2$, $c = K_1 K_2 K_3$, $d = K_1 K_2 K_3 K_4$, and x, y, z, and w are the terms in $\{H^+\}$ and F. For example,

$$y = \frac{\{H^+\}^2(2 - F)}{(4 - F)}.$$

Equation (3.26) can be solved by the method of least squares to yield

$$a\Sigma x^2 + b\Sigma x \cdot y + c\Sigma x \cdot z + d\Sigma x = \Sigma x \cdot w$$
$$a\Sigma x \cdot y + b\Sigma y^2 + c\Sigma y \cdot z + d\Sigma y = \Sigma y \cdot w$$
$$a\Sigma x \cdot z + b\Sigma y \cdot z + c\Sigma z^2 + d\Sigma z = \Sigma z \cdot w$$
$$a\Sigma x + b\Sigma y + c\Sigma z + d \cdot n = \Sigma w$$

where n is the total number of experimental observations. This set of equations can be solved for the constants by a method for which computer subroutines are available (see for example Dickson, 1968). An alternative solution of equation (3.26) is by an iteration method for which purpose it can be rearranged to

$$\frac{1}{K_1} = \frac{(1 - F)}{\{H^+\} \cdot F} + K_2 \frac{(2 - F)}{\{H^+\}^2 \cdot F} + K_2 K_3 \frac{(3 - F)}{\{H^+\}^3 \cdot F} + K_2 K_3 K_4 \frac{(4 - F)}{\{H^+\}^4 \cdot F} \tag{3.27}$$

which can be written as

$$\frac{1}{K_1} = A_1 + K_2 A_2 + K_2 K_3 A_3 + K_2 K_3 K_4 A_4 \ldots \tag{3.27a}$$

where A_1, A_2, A_3, A_4 correspond to the terms in $\{H^+\}$ and F in equation (3.27). The continuation dots emphasize that the equation could be extended in a logical sequence. To initialize the procedure, all terms other than A, on the right-hand side of equation (3.27a) are ignored and the results for the first equivalent used to obtain an approximation of K_1. The equation is rearranged to yield values of $1/K_2$ and is solved using the mean value of K_1 in conjunction with experimental values of A_1 and A_2 calculated from the second equivalent values. The approximate mean value of K_2 is used to refine the values of K_1 for the first equivalent results by introducing the term $K_2 \cdot A_2$ in equation (3.27a). Similarly the approximate value of K_3 is calculated from the third equivalent and K_4 from the fourth equivalent. The calculation continues until successive values of K_1, K_2, K_3, and K_4 are invariant when the whole equation is used. This method of calculation lent itself readily to our computer program, POLY 3, which can provide as many as six overlapping constants.

These mixed constants are used, in turn, to calculate the contribution of the anionic species to the ionic strength (see equation 3.24 and also Pecsok and Shields, 1976). The ionic strength is calculated by an equation analogous to (3.25) remembering that the squares of the valency terms are required for this purpose. The activity coefficients are calculated and used only to obtain the hydrogen ion concentration to refine the values of F (see equation 3.22a). The whole calculation is then repeated until successive values of the mixed constants are the same.

We wish to make it clear that these constants may have little significance except for calculating the ionic composition of the solution at a given pH, or for subsequent application to determine the stability constants of metal complexes which the compound may form. This is because they are 'macroscopic constants' which cannot be related to the ionizing property of a particular group without further experimental work (see p. 130 and also King, 1965, p. 227). An additional restriction is placed upon the interpretation of the pH-titrations of polymeric acids (e.g. polyacrylic acid) or ampholytes (e.g. proteins) unless these (as rarely) were isolated as homomolecular entities; otherwise such data cannot be interpreted even in terms of macroscopic constants. Instead, wrongly assuming that each functional group will ionize identically and independently, titration curves are interpreted in terms of either a single microscopic constant for a polymeric acid or according to the classes of functional groups in the case of proteins. Deviations from this assumption are invariably encountered and are associated with electrostatic interactions between the groups. A detailed discussion of the nature of these interactions is beyond the scope of this book and the reader is referred to a review by Doty and Ehrlich (1952), and to King (1965, p. 230) where more information is to be found.

3.10 Accuracy of the potentiometric method

The pH-titration method is a stringent test of the performance of the glass electrode. This must retain its initial standardization over a pH range in which the ionic strength is changing and which is often far removed from that of the standardizing buffer solution. Other effects, although compensated in the initial standardization, may change as the titration proceeds, e.g. the junction potential. In view of the factors, the question must be asked: how accurately can pK_a values be determined using pH meters?

When using the routine technique described in Chapter 2 coupled with the refinements of calculations described in this chapter, the results are likely to be within $0.04 pK$ of the true thermodynamic value for the range pK 2–10 determined at concentrations in the range 0.0025–0.050M. For example, the results for benzoic acid, shown in Table 2.2, when refined by the calculations outlined above, gave the thermodynamic pK_a 4.20 ± 0.02 (cf. the accepted figures of 4.18–4.20 as obtained by the most refined and painstaking methods, Kortüm *et al.*, 1961). By changing to a pH meter calibrated in 0.002 pH divisions, the precision of the measurements of pH can be increased by a factor of 5, without necessarily increasing the accuracy by a similar factor, even though the refinements of technique given on p. 41 are rigidly employed. The excellent results recorded in Table 3.3 prompted us to study in more detail the likely accuracy of the method, particularly in view of the increasing use of pH meters having a similar precision. Hence twenty-eight titrations, using succinic acid as the trial substance, at a series of concentrations ranging from 0.001 to 0.025M, were performed in quadruplicate. Two operators took part; one with, and one without previous experience. Twenty-six readings of pH corresponding to the degrees of neutralization given in Table 3.3 were obtained for each titration. The overall spread (of each pH reading) from the mean value for each concentration was noted. This spread was found to differ significantly with the concentration. At concentrations of 0.0025M and below, no good duplication was observed between any two titrations out of the four. As the concentration was increased, the replication became better, as demonstrated by the mean spread of 104 pH values obtained at each concentration: ± 0.014 at 0.0025M, ± 0.006 at 0.0050M, ± 0.004 at 0.0100M, and ± 0.005 at 0.0250M. In the last three concentrations, good duplication was observed for titrations within a given concentration and no correlation could be found between the best duplicates and operator experience. Treating the results obtained at 0.005, 0.01 and 0.025M as a composite set of data, yielded 156 values both for pK_1 and pK_2. The mean of these were $pK_1 = 4.206$ and $pK_2 = 5.635$, which agree quite well with those published by Bates. At concentrations of 0.0025M and below, however, the pK_a values could be precise but wrong. At 0.0025M, for example, the following values were obtained: $pK_1 = 4.199 \pm 0.003$, 4.184 ± 0.012, 4.186 ± 0.005, 4.173 ± 0.007; and pK_2 values of 5.608 ± 0.028, 5.643 ± 0.017, 5.627 ± 0.004, 5.618 ± 0.003. Repeating these titrations in the presence of 0.05M potassium chloride, yielded results which were just as erratic, although the time taken to reach an equilibrium pH value after addition of titrant (see p. 42) was much reduced in the presence of the salt.

These results imply that, in addition to the limitation of the pH range over which the glass electrode is thought to be accurate, there is also a limitation upon the *concentration* range over which its use is applicable for the accurate determination of pK_a values. This conclusion has been supported by the work of Serjeant and Warner (1978) who used the glass electrode and a silver–silver chloride electrode in a cell without liquid junction in order to assess the accuracy of the glass electrode. The measured potentials of this cell were used to calculate the thermodynamically valid acidity function, $p(a_H \gamma_{Cl})$ (see Appendix VI) for buffer solutions containing equimolal amounts of potassium dihydrogen phosphate and disodium hydrogen phosphate plus known concentrations of chloride ion. The maximum error assessed for a single measurement in a given solution increased with the dilution and was found to be ± 0.003, ± 0.003, and ± 0.008 $p(a_H \gamma_{Cl})$ unit for the respective phosphate buffer concentrations of 0.02, 0.0025, and 0.001 m. At a concentration $\geqslant 0.0025$ m, the accuracy approached that of the hydrogen electrode for these measurements. The overall error must, of course, be greater in the measurement of pH because the uncertainty as to the magnitude of the liquid junction potential. A method has been developed for the determination of pK_a values by using $p(a_H \gamma_{Cl})$ titrations in preference to the pH titrations described in this chapter. Experimental details, together with worked examples are given by Serjeant (1983) and good accuracy is reported. The advantage of using this method is that the pK_a values derived from these titrations have strict thermodynamic validity because the cells contain no liquid junction. However, the possible reactivity of the silver–silver chloride electrode with amines (which are frequent candidates for pK_a measurement) imposes a restriction on the use of the method, hence it has not been included here.

The accuracy of the pH titration method, therefore, is such that the use of precision pH apparatus for the determination of the pK_a values of sparingly soluble substances offers no advantage over the use of pH meters calibrated in 0.01 (or even 0.02) pH unit. However, in the concentration range 0.005–0.050M, the results obtainable by use of precision pH apparatus can closely approach the true thermodynamic value within the range pK 3–9.5, if titrations are replicated and great care is exercised in the standardization procedure. Outside this restricted range, uncertainties are introduced by (a) the magnitude of the correction terms used in the calculations, and (b) significant departures from accuracy in the glass electrode above pH 10.

3.11 Non-aqueous solvents

The order of relative strengths of a series of bases in anhydrous acetic acid has been shown to parallel their order of strengths in water. This result induced Hall (1930) to measure in acetic acid a number of those very weak bases which are too readily hydrolysed to be titrated potentiometrically in water. To the pK_a values were added 2.0, which was the average difference for those bases that could be titrated both in acetic acid and in water. The results tended to be inexact (e.g. pK_a + 0.06 instead of − 0.29 for *o*-nitroaniline at 25°C), but

Table 3.4

```
 1 C        PROGRAM PKDI
 2 C        CALCULATES OVERLAPPING PK VALUES OF
 3 C        BIFUNCTIONAL COMPOUNDS FROM POTENTIOMETRIC
 4 C        DATA AT 20 OR 25C.
 5 C
 6          COMMON PH(50),HACT(50),H(50),POH(50),OH(50),
 7         1CONCA(50),VOLA(50),CONCS(50),ACT(50),ACT1(50),
 8         2ACT2(50),F(50),A(50),B(50),X(50),Y(50),DENOM(50),
 9         3HA(50),AA(50),STREN(50),FS(50),HCONC(50),FK1(50),
10         4PK1(50),FK2(50),PK2(50),NSUBS(40),H2A(50),CL(50)
11 C
12 C
13 C        READ THE TOTAL NUMBER OF RESULTS (N) AND THE NUMBER IN
14 C        THE FIRST EQUIVALENT (K). READ THE CONSTANT VALUES
15 C        IN THE DATA. THESE ARE THE MOLECULAR WEIGHT (SMOL)
16 C        THE WEIGHT DISSOLVED IN THE INITIAL VOLUME(AVOL), THE
17 C        CONCENTRATION OF TITRANT (ACONC). FOR DIACIDIC BASES
18 C        AND AMPHOLYTES READ THE VOLUME OF ACID, (VOL) AND ITS
19 C        CONCENTRATION (ACID) USED IN THE INITIAL DISSOLUTION.
20 C        READ THE TEMPERATURE (T) AND THE TYPE OF COMPOUND. (KTYPE) WHICH
21 C        IS ALLOCATED AS FOLLOWS. NEGATIVE INTEGER FOR ACIDS,
22 C                                 ZERO FOR AMPHOLYTES,
23 C                                 POSITIVE INTEGER FOR BASES.
24 C
25 C        READ THE NAME OF THE SUBSTANCE (NSUBS)
26     986 READ(3,20) N,K,SMOL,AVOL,WT,ACONC,T,KTYPE,
27         1ACID,VOL,(NSUBS(I),I = 1,40)
28      20 FORMAT (2I3,2F10·2,F10·5,F10·4,F5·1,I3,
29         1F10·4,F10·2,/,40A2)
30         IF (N) 999,999,99
31      99 WRITE(4,21) (NSUBS(I),I = 1,40)
32      21 FORMAT(1H1,///,30X,10HSUBSTANCE- ,40A2)
33         WRITE (4,39) SMOL,WT,AVOL,VOL,ACID,ACONC,T
34      39 FORMAT (32X,17HMOLECULAR WEIGHT  = ,F5·1,//,24X,
35         1F7·5,15HG DISSOLVED IN ,F6·2,11HML OF WATER,
36         2/,26X,11HCONTAINING  ,F5·2 ,6HML OF ,F6·4,
37         36HM HCL  ,/,10X,17HMOLARITY TITRANT= ,F6·4,1HM,31X,
38         412HTEMPERATURE= ,F4·1///11X,3HVOL,4X,2HPH,9X,
39         51HX,13X,1HY,13X,1HI,10X,3HPK1,4X,3HPK2)
40 C
41 C        READ N VALUES OF THE VOLUME OF TITRANT
42 C        AND PH
43 C
44         READ (3,22) (PH(I),VOLA(I),I = 1,N)
45      22 FORMAT(2F10·4)
46         DO 1 I = 1,N
```

Table 3.4 (*contd.*)

```
47            IF (T.EQ.25·0) GO TO 23
48            IF (T.EQ.20·0) GO TO 24
49       23 POH(I) = 14·000 − PH(I)
50            GO TO 25
51       24 POH(I) = 14·160 − PH(I)
52       25 H(I) = 10·0∗∗(− PH(I))
53            HACT(I) = H(I)
54            OH(I) = 10·0∗∗(− POH(I))
55            CONCA(I) = VOLA(I)∗ACONC/(AVOL + VOLA(I))
56            CONCS(I) = WT∗10·0∗∗3/(SMOL∗(AVOL + VOLA(I)))
57            ACT1(I) = 1·0
58            ACT2(I) = 1·0
59        1 ACT(I) = 1·0
60 C
61 C          SOLVE EQUATIONS (3·22) AND (3·23) BY LEAST SQUARES TO
62 C          OBTAIN THE MIXED CONSTANTS, CK1 AND CK2 AND THE
63 C          THERMODYNAMIC CONSTANTS, TK1 AND TK2. INITIALLY THESE
64 C          TWO SETS OF CONSTANTS WILL BE EQUAL.
65 C
66            TK1 = 0·0
67            TK2 = 0·0
68       11 SUMX = 0·0
69            SUMY = 0·0
70            SUMXY = 0·0
71            SUMX2 = 0·0
72            SUMA = 0·0
73            SUMB = 0·0
74            SUMAB = 0·0
75            SUMA2 = 0·0
76            TK1A = TK1
77            TK2A = TK2
78            DO 2 I = 1,N
79            F(I) = (CONCA(I) + HACT(I) − OH(I))/CONCS(I)
80            A(I) = H(I)∗∗2∗F(I)/(2·0 − F(I))
81            B(I) = H(I)∗(1·0 − F(I))/(2·0 − F(I))
82            IF(KTYPE) 26,27,28
83       26 X(I) = A(I)/ACT2(I)
84            Y(I) = B(I)/ACT1(I)
85            GO TO 29
86       27 X(I) = A(I)
87            Y(I) = (B(I)/ACT(I))
88            GO TO 29
89       28 X(I) = A(I)∗ACT2(I)
90            Y(I) = B(I)∗ACT(I)
91       29 SUMY = SUMY + Y(I)
92            SUMB = SUMB + B(I)
93            SUMX = SUMX + X(I)
```

Table 3.4 (*contd.*)

```
94        SUMA = SUMA + A(I)
95        SUMXY = SUMXY + X(I)*Y(I)
96        SUMAB = SUMAB + A(I)*B(I)
97        SUMX2 = SUMX2 + X(I)**2
98      2 SUMA2 = SUMA2 + A(I)**2
99        FN = N
100       DENOM1 = (FN*SUMX2 – SUMX**2)
101       DENOM2 = (FN*SUMA2 – SUMA**2)
102       SLOPE1 = (FN*SUMXY – SUMX*SUMY)/DENOM1
103       SLOPE2 = (FN*SUMAB – SUMA*SUMB)/DENOM2
104       CEPT1 = (SUMX2*SUMY – SUMX*SUMXY)/DENOM1
105       CEPT2 = (SUMA2*SUMB – SUMAB*SUMA)/DENOM2
106       TK1 = 1·0/SLOPE1
107       CK1 = 1·0/SLOPE2
108       TK2 = ABS(CEPT1)
109       CK2 = ABS(CEPT2)
110 C
111 C     COMPUTE IONIC STRENGTH AND ACTIVITY FUNCTIONS
112 C     AT THE APPROPRIATE TEMPERATURES FOR THE GIVEN
113 C     TYPE OF COMPOUND.
114 C
115       DO 3 I = 1,N
116       DENOM(I) = (H(I)**2 + CK1*H(I) + CK1*CK2)
117       HA(I) = ((CK1*H(I))/DENOM(I))*CONCS(I)
118       AA(I) = ((CK1*CK2)/DENOM(I))*CONCS(I)
119       H2A(I) = (H(I)**2/DENOM(I))*CONCS(I)
120       CL(I) = VOL*ACID/(AVOL + VOLA(I))
121       IF (KTYPE) 36,37,38
122    36 STREN(I) = 0·5*(CONCA(I) + HA(I) + 4·0*AA(I)) + 0·5*HACT(I)
123       GO TO 35
124    37 STREN(I) = 0·5*(CL(I) + H2A(I) + CONCA(I) + AA(I)) + 0·5*HACT(I)
125       GO TO 35
126    38 STREN(I) = 0·5*(CL(I) + 4·0*H2A(I) + CONCA(I) + HA(I)) + 0·5*HACT(I)
127    35 FS(I) = STREN(I)**0·5/(1·0 + 1·5*(STREN(I)**0·5))
128       IF (T.EQ.25·0) GO TO 34
129       IF (T.EQ.20·0) GO TO 33
130    34 D = 0·5115
131       GO TO 32
132    33 D = 0·5070
133    32 ACT(I) = 10·0**(D*FS(I))
134       ACT(I) = 10·0**(3·0*D*FS(I))
135       ACT2(I) = 10·0**(4·0*D*FS(I))
136       HCONC(I) = H(I)*ACT(I)
137       HACT(I) = HCONC(I)
138     3 OH(I) = 10·0**(– 14·00)/HACT(I)
139 C
140 C     CHECK CONVERGENCE OF SUCCESSIVE VALUES
```

Table 3.4 (*contd.*)

```
141  C       FOR EACH CONSTANT
142  C
143          EPSILA = ABS(TK1*10·0**(−5))
144          EPSILB = ABS(TK2*10·0**(−5))
145          DIFFA = ABS(TK1A − TK1)
146          DIFFB = ABS(TK2A − TK2)
147          IF (DIFFA.GT.EPSILA) GO TO 11
148          IF (DIFFB.GT.EPSILB) GO TO 11
149  C
150  C       CALCULATE PK1 AND PK2
151  C
152          SUM = 0·0
153          DO 4 I = 1,K
154          FK1(I) = X(I)/(Y(I) + TK2)
155          PK1(I) = ALOG10(1·0/FK1(I))
156        4 SUM = SUM + PK1(I)
157          FN = K
158          AV1 = SUM/FN
159          WRITE (4,112)(VOLA(I),PH(I),X(I),Y(I),STREN
160        1(I),PK1(I),I = 1,K)
161      112 FORMAT(10X,F5·3,F7·3,3E14·5,F9·3)
162          SUM = 0·0
163          K1 = K + 1
164          DO 5 I = K1,N
165          FK2(I) = SLOPE1*X(I) − Y(I)
166          PK2(I) = ALOG10(1·0/FK2(I))
167        5 SUM = SUM + PK2(I)
168          FN = N − K
169          AV2 = SUM/FN
170          WRITE(4,113)(VOLA(I),PH(I),X(I),Y(I),STREN
171        1(I),PK2(I),I = K1,N)
172      113 FORMAT(10X,F5·3,F7·3,3E14·5,9X,F7·3)
173          WRITE(4,132)AV1,AV2
174      132 FORMAT(//,64X,12HAVERAGE PK1 = ,F5·3,/,
175        172X,4HPK2 = ,F5·3)
176          GO TO 986
177      999 CALL EXIT
178          END
```

they served to open up a territory which might otherwise remain unexplored, namely weak bases lacking ultraviolet spectra (e.g. urea, pK_a + 0·10).

More recently, methods have been developed for the determination of ionization constants not only in acetic acid, but also in acetonitrile, dimethylformamide, N-methyl-2-pyrrolidone, dimethylsulphoxide, 2-methyl-2-propanol, and some ketones. The methods used for the standardization of cells containing glass electrodes and the subsequent determination of pK values in these solvents are reviewed by Serjeant (1984).

4 Determination of ionization constants by spectrophotometry

4.1 Introduction

The determination of ionization constants by ultraviolet or visible spectro-photometry is more time-consuming than by potentiometry. However, spectro-metry is an ideal method when a substance is too insoluble for potentiometry or when its pK_a value is particularly low or high (e.g. less than 2 or more than 11). The method depends upon the direct determination of the ratio of molecular species (neutral molecule) to ionized species in a series of non-absorbing buffer solutions (whose pH values are either known or measured). For this purpose, the spectrum of the *molecular species* must first be obtained in a buffer solution whose pH is so chosen that the compound to be measured is present wholly as this species. This spectrum is compared with that of the pure *ionized species* similarly isolated at another suitable pH. A wavelength is chosen at which the greatest difference between the absorbances of the two species is observed. This is termed the 'analytical wavelength'. Fig. 4.1, in which the base 2-aminopyridine is used as an example, shows how these two pH values can be found.

Fig. 4.1 shows a series of spectra, in buffers spaced at intervals of 1 pH unit, collected on one chart (recording spectrophotometer). The data in Fig. 4.1 suggested that the spectrum of the molecular species be measured at pH 10 and that of the cation at pH 4. The greatest difference between their absorbances was seen to be at 310 nm, and hence this was selected as the analytical wavelength. By using this at various pH values, intermediate between those at which the spectra of the two species were obtained, the ratio of ionized to molecular species can be calculated. This is possible because a series of two-component mixtures is formed in which the ratio of the two species depends solely upon the pH at which the solution is optically measured. If it is assumed that Beer's Law is obeyed for both species (and this is usually the case), the observed absorbance (A) at the analytical wavelength is the sum of the absorbances of the ionized species (A_I) and the molecular species A_M), thus:

$$A = A_I + A_M.$$

The absorbance of either component is related to its molar concentration (c) by a general expression ($A = \varepsilon l c$), where ε is the molar absorption coefficient of the particular species and l is the optical length of the cell.

The concentration of the ionized species in the mixture is $F_I c$, where F_I is the fraction ionized, thus (for acids):

$$F_I = [A^-]/([A^-] + [HA]).$$

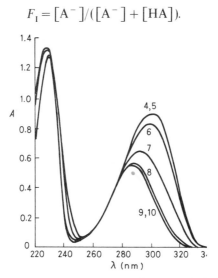

Figure 4.1 Example of graph produced in searching for the analytical wavelength. The test substance is a base, 2-aminopyridine. The spectrum of the neutral species is seen at pH 9 and 10, that of the cation at pH 4 and 5.

For an analytical wavelength, 310 nm was selected and, using it, pK_a 6·82 was found.

Hence the contribution of F_I to the observed absorbance of the mixture is $\varepsilon_I F_I cl$. Similarly, the contribution of the molecular species to the observed absorbance of the mixture is $\varepsilon_M F_M cl$, where F_M is the fraction present in the molecular form. The terms ε_I and ε_M are the molar absorption coefficients of the ionized and molecular species, respectively, which are related directly to the absorbances obtained at pH 4 and pH 10 in the example (Fig. 4.1). The fractions of ionized and molecular species (see p. 45) are given by expressions obtained by replacing $[HA]$ by $[H^+][A^-]/K_a$:

$$F_I = K_a/([H^+] + K_a)$$

and
$$F_M = [H^+]/([H^+] + K_a) \quad \text{for acids}$$

and by

$$F_I = [H^+]/([H^+] + K_a)$$
and
$$F_M = K_a/([H^+] + K_a) \quad \text{for bases.}$$

If the same cell optical length is used throughout, then

$$A = (\varepsilon_I \cdot F_I + \varepsilon_M \cdot F_M)c$$

and therefore, because $\varepsilon = A/c$, and F_I and F_M are defined above,

$$\varepsilon = \frac{\varepsilon_I \cdot K_a}{[H^+] + K_a} + \frac{\varepsilon_M [H^+]}{[H^+] + K_a} \quad \text{for acids.} \tag{4.1a}$$

Similarly

$$\varepsilon = \frac{\varepsilon_M \cdot K_a}{[H^+] + K_a} + \frac{\varepsilon_I [H^+]}{[H^+] + K_a} \quad \text{for bases.} \tag{4.1b}$$

Provided that the same total concentration is used for all measurements, equations (4.1a) and (4.1b) may be written with the appropriate absorbance (A) replacing the molar absorption coefficient (ε). Each equation can be arranged in two forms to suit different needs.

When the functional group being determined is an acid, equation (4.2a) is used if A_I is greater than A_M, whereas (4.2b) is used if the reverse is the case.

$$pK_a = pH + \log \frac{A_I - A}{A - A_M} \tag{4.2a}$$

$$pK_a = pH + \log \frac{A - A_I}{A_M - A} \tag{4.2b}$$

When the group being determined is a base, equation (4.3a) is used if A_I is greater than A_M, and (4.3b) if the reverse is the case.

$$pK_a = pH + \log \frac{A - A_M}{A_I - A} \tag{4.3a}$$

$$pK_a = pH + \log \frac{A_M - A}{A - A_I} \qquad (4.3b)$$

The use of these equations is demonstrated by the worked examples given in Tables 4.3, 4.4a, and 4.5 which should be consulted at this stage. Below we discuss the general procedure and practical factors which must be taken into account when using this method. The operations required can be summarized as

(a) the preparation of the stock solution, and suitable dilutions of it in appropriate buffers,

(b) the search for the pure spectra of the two species involved in the equilibrium,

(c) the choice of an analytical wavelength suitable for the determination,

(d) the search for an approximate value of the pK_a using one 'sighting' reading,

(e) the exact determination of the pK_a using at least seven results.

4.2 Apparatus

The apparatus required consists of a photoelectric spectrophotometer, recently recalibrated (for wavelength *and* optical density) and free from stray light above 220 nm; 1- and 4-cm matched cells (quartz with covers); graduated flasks and pipettes. An instrument for the potentiometric determination of pH must be at hand.

Manually operated spectrophotometers are particularly suitable for this type of work. Most recording instruments do not reproduce density readings accurately enough for determination of pK, but can help in shortening the time taken to find the analytical wavelength.

4.3 Buffers

Buffers are aqueous solutions devised to resist pH change on addition of acid or alkali, or upon dilution. Only those with but little optical absorbance are suitable for use in spectrophotometry. A selection of substances suitable for preparing such buffers is given in Table 4.1. Details are available for converting them into buffers at a low (and constant) ionic strength (I) for which 0·01 is a suitable value (Perrin, 1963; Perrin and Dempsey, 1974, p. 32).

These solutions are conveniently stored at 0·1M strength in glass-distilled water. For use, they are diluted to 0·01M and adjusted with N-potassium hydroxide, or hydrochloric acid, to the required pH. No solutions should be stored in polythene containers, which usually liberate an optically absorbing plasticizer. Usually about 100 ml 0·01M buffer solution is required for each

Table 4.1   Substances for buffer solutions

These substances, when partly neutralized, form buffers suitable for spectroscopic determination of ionization constants.* (Perrin and Dempsey, 1974)

Substance	Useful pH range
Chloroacetic acid	2·2–3·4
Formic acid	3·2–4·4
Acetic acid	4·1–5·3
Succinic acid	3·6–6·2
Citric acid	5·8–6·7
Phosphoric acid	6·4–7·7
Ethylenediamine monohydrate	6·6–7·8 (also 9·3–10·5)
N-Ethylmorpholine	7·0–8·2
Trishydroxymethylaminomethane ('Tris')	7·7–8·9
Boric acid – borax	8·5–9·7
Bicarbonate	9·6–10·6
Butylamine	10·3–11·6

*Some sighting values may be derived from Table 2.1 (p. 18)

spectroscopic measurement. If buffers other than those in Table 4.1 are used, they should first be tested to make sure that they are not significantly more absorbent in the region under investigation. The following may be useful when bases form a precipitate with the anions in other buffers:

(a) N-Ethylmorpholine, b.p. 138–139°C, capacity 7·0–8·2. It should be fractionally distilled before use;
(b) Ethylene diamine monohydrate, b.p. 118°C.

The *buffer capacity* of a solution is defined as the amount of strong acid or alkali required to produce unit change of pH in the solution. The buffer capacity, of solutions made from substances in Table 4.1, is greatest near the middle of the given pH ranges, and very slight at the ends of these ranges. When working at a pH that falls between the capacity of two consecutive buffers, the solution should be made 0·01M with respect to *both* buffers. Thus, a mixture of acetic acid and potassium dihydrogen phosphate (0·01M, in each component) forms an efficient buffer for the range 5·2–6·5.

Because the optical absorption of these solutions, although low, is not negligible, it is important to get precisely the same proportions (of buffers to other aqueous solutions) in both optical cells. For example, if the first dilution of the unknown consists of 98 parts buffer and 2 parts aqueous stock solution, the comparison cell should contain a solution of buffer similarly diluted (98:2) with water.

It is also important to note that the absorption of the solutions changes with the pH. Hence the pH of the contents of both optical cells must be identical to 0·03 unit, and it is advisable to check this once more after the absorbance has been measured.

For pH values outside the range given in Table 4.1, hydrochloric acid and sodium or potassium hydroxide are used (see the concentration columns in Tables 11.4 and 11.5 of Appendix III). This is because water has an acidic pK_a of about 14 and a basic pK_a about 0, at 25°C. Hence it begins to develop some buffer capacity below pH 2 and above 12.

Because quartz cells are etched by alkali, solutions of pH 13 should not be left in them for more than half an hour. Solutions of higher pH are best used in glass cells, at such wavelengths as glass transmits; but if quartz cells have to be used at pH > 13, they must be used quickly and rinsed immediately.

4.4 Acidity functions

(Readers who are not concerned with pK_a values lying outside the range 1–13 should pass on to the next section, p. 78.)

To provide a series of buffers, e.g. for the optical determination of pK_a values, in highly acid regions, Hammett and Deyrup (1932) devised the acidity function, denoted H_0 (see also Hammett, 1970). Thus for determining the ionization constants of very weak bases, solutions of known H_0 take the place of the buffer solutions mentioned above. It will be noted that H_0 values, although not preceded by the symbol p are negative logarithms, and that they form an extension of the pH scale. The H_0 values of the solutions in Table 4.2 are from the results of Bascombe and Bell (1959) up to 40 wt% H_2SO_4; beyond this strength those of Ryabova, Medvetskaya and Vinnik, (1966) have been used, a pair of choices endorsed by Rochester (1970). Above 30°C, the values decrease a little with further rise in temperature (Tickle, Briggs and Wilson, 1970).

The H_0 scale was constructed by measuring the ratio $[BH^+]/[B]$ for *p*-nitroaniline (the pK_a of which is known) in solutions of increasing acidity. When this ratio became high, a less basic indicator was substituted; its pK_a was calculated by measuring the $[BH^+]/[B]$ ratio in a solution of which the H_0 had been determined with *p*-nitroaniline, and similar measurements (with overlapping indicators, in solutions of ever-increasing acidity) finally enabled the calibrated scale to extend to $H_0 - 13$.

Although not apparent in 1932, it is now known that, to obtain a smooth curve and significant relationship, all the indicators used in constructing an H_0 scale should be primary amines, as in the data given in Table 4.2. Some authors prefer to write pK_{BH^+} instead of pK_a for results obtained with H_0 solutions. This indicates that an acidity function has been used; also it distinguishes the result from those obtained in deuterated acids which are base-strengthening.

Table 4.2 Solutions of known acidity function (H_0) at 25°C

H_2SO_4 (wt%)	Normality	H_0
0·0049	0·001	3·03
0·047	0·01	2·08
0·612	0·13	1·06
5·10	1·08	0·09
8·30	1·80	− 0·23
10·75	2·36	− 0·42
11·8	3·08	− 0·64
17·8	4·08	− 0·90
22·2	5·22	− 1·21
25·1	6·04	− 1·41
28·7	7·10	− 1·69
33·6	8·58	− 2·03
35·6	9·22	− 2·22
38·6	10·18	− 2·45
42		− 2·69
47		− 3·13
52		− 3·60
60		− 4·51
70		− 5·92
80		− 7·52
85		− 8·29
90		− 9·03
99		− 10·57
100		− 11·94

Thus 2, 4-dinitroaniline has pK_{BH^+} − 4·42, whereas its pK_{BD^+} is − 4·03 (Högfeldt and Bigeleisen, 1960). It is often asked why acidity, as measured by H_0, increases so much faster than the stoichiometric concentration of acid rises? Apparently it is because the hydration of the hydrogen ion declines as the concentration of water falls.

The H_0 scale is valuable for determining the pK_a values of all weak *primary* amines and heteroaromatic amines such as pyridine (Johnson *et al.*, 1965) and its azalogues (e.g. pyrimidine). When used for other substances it is prudent to plot H_0 against the absorption coefficient (ε). If a sigmoid curve resembling Fig. 1.1 (within an interval of about 2.5 H_0 units) is not obtained, a different acidity function should be tried.

It is obvious that water becomes decreasingly available as a solvent as the acidic strength of solutions rises. This can lead to 'medium shifts', a common source of error, which can be guarded against in the following way. The spectrum of the cation should be recorded at a H_0 two units more negative than the supposed pK_a (at such a H_0, less than 1% of neutral species should be present). This spectrum should be identical with spectra found in stronger solutions of the same acid: if it is not, a medium effect is present and the determined pK_a may prove to be seriously in error (Noyce and Jorgenson, 1962).

Tertiary benzoaromatic bases such as NN-dimethylaniline and N-methyldiphenylamine have different activity coefficients, possibly because hydration of the cation is diminished by steric influences. Arnett and Mach (1964) specified a set of sulphuric acid solutions to give an appropriate acidity function H_0'''. For comparison: 40 wt% H_2SO_4 gives -3.46 on this 'triple prime' scale, but only -2.54 on Hammett's scale (H_0).

A still steeper slope, in the plot of acidity function against sulphuric acid concentration, is shown by carbenium ions, which are yet poorer hydraters. Straightforward examples include naphthalene, diarylethylenes, azulenes, and 1, 3, 5-trimethoxybenzene (the latter protonates on C-2). For these, Reagen (1969) devised an H_C acidity function. However, if hydrogen-bonding groups (such as amino- or hydroxy) were added to these nuclei, the acidity function moved strongly to H_0.

The J_0 (also called H_R) functions devised for triphenylcarbinols are similar to H_C. The carbinols protonate on oxygen and then eliminate a molecule of water to give a carbenium ion. The steep slope of the acidity function curve is exemplified by 40 wt% H_2SO_4 which gives -4.80 on the J_0 scale but only -2.54 for H_0 (Arnett and Bushick, 1965; Murray and Williams, 1950). Amino-derivatives of triphenylcarbinols follow the H_0''' scale (Belotserkovskaya and Ginsberg, 1964).

Indoles protonate on C-3 and pyrroles on C-2 (Hinman and Lang, 1964, and Chiang and Whipple, 1963, respectively). Hinman's acidity function (H_I) for these substances is numerically very close to H_0'''. Ketones, of which substituted benzophenones are well explored, have been given the function H_B the slope of which is identical with than of H_0 up to 60 wt% H_2SO_4, above which a much steeper slope sets in (Bonner and Phillips, 1966).

Amides, which protonate on the oxygen atom (Fraenkel and Franconi, 1960), produce cations which are much more strongly hydrated than those of primary amines. Hence the slope of the characteristic function, H_A, is much flatter than that of H_0 (Katritzky, Waring and Yates, 1963; Yates, Stevens and Katritzky, 1964). Pyridin-2-one (formerly called 2-hydroxypyridine) and uracil (pyrimidin-2, 4-dione) and their derivatives all follow the H_A scale (Bellingham, Johnson and Katritzky, 1967; Katritzky and Waring, 1962).

So far, we have dealt only with uncharged bases and their conversion to cations, and all acidity functions discussed have been variants of Hammett's H_0, the subscript of which signifies zero charge. Let us briefly discuss some other cases. The function H_+ has been devised for the monocations of bases whose second basic pK_a is under investigation. This function has been found to resemble H_0 for the aminopyridine cations and H_A for the second protonation of pyrimidine and pyrazine (Brignell *et al.*, 1967). For the protonation of the anion of strong acids (which yields the molecule), the function H_- is used (Paul and Long, 1957). This scale was constructed by the use of optically absorbing acids just as bases were used for the H_0 scale. There are indications that a whole set of H_- functions may be needed to deal with acids of different constitutions (Reeves, 1966) and this remark applies to H_+ functions also. Ionization constants of the following strong acids were found with the help of H_- acidity functions: methanesulphonic acid, *p*-toluenesulphonic acid, and nitric acid (Covington and Lilley, 1967; Dinius and Choppin, 1962; and Hood and Reilly, 1960, respectively). In all three cases, proton magnetic resonance was used in place of spectrophotometry as the measuring technique, because of insufficient differences in λ_{max} (see Chapter 7).

It is evident that yet other acidity functions could exist at the extreme alkaline edge of the pH range, namely above pH 14, for measuring the pK_a values of weak acids and strong bases, the former with an H_- scale and the latter with an H_0 scale. This is a more difficult region of pH than the acidic stretch dealt with in the foregoing, in so far as the glass electrode becomes increasingly inaccurate, and strong OH^- absorption swamps the readings. It is well established that the basic properties of aqueous alkalies increase, in a non-linear fashion, with concentration (Schwarzenbach and Sulzberger, 1944). For examples of the use of H_- in highly alkaline solution, see Rochester (1966, 1970), Jones (1973). In general it may be said that the determination of ionization constants in highly alkaline solution is a comparatively unexplored territory. Here, as in every application of acidity functions to untypical substances, the sigmoid curve test (see above) should be carried out to make sure that the function being used is a relevant one.

For further reading on acidity functions, see Rochester's book (1970).

After this excursion into the territory of acidity functions (possibly more interesting to the physical organic chemist than to the biological chemist) we resume the main narrative.

4.5 Preparation of the stock solution of the unknown

If solubility permits and the absorption coefficient is not unusually low, a convenient strength for a stock solution is 5×10^{-4}M. (However, substances of high absorption coefficient have given pKs with a precision of ± 0.02 when poor solubility dictated that stock solutions as weak as 2×10^{-5}M be used.) To facilitate dissolution, an acidic compound may be dissolved in 0.005N potassium hydroxide (if the pK_a is believed to lie below 10), and conversely a basic compound may be dissolved in 0.005N hydrochloric acid (if the pK is believed to lie above 4). The good wetting properties of alcohol sometimes indicate its use for a concentrated primary solution, but the alcohol content must not exceed 1% in the final stock solution.

4.6 The search for the spectra of two pure ionic species

If the unknown is an acid, two spectra are required: that of the anion and that of the molecule. The stock solution is diluted to 10^{-4} in 0·01N hydrochloric acid and 0·01N potassium hydroxide (i.e. at approximately pH 2 and 12, respectively). The absorbance is then measured over the whole spectrum. The spectrum obtained in acid is that of the molecule and the other is that of the anion. To see if either of these is an impure spectrum (i.e. if it contains the other ionic species) the unknown is again examined in 0·1N hydrochloric acid and 0·1N potassium hydroxide (i.e. at pH 1 and 13). If there is no change greater than 1% in the absorbance of any peak, the two species may be considered to be isolated. If there is a change, the measurements should be repeated yet one pH unit further from neutrality.

The possibility of the following irregularities should be borne in mind:

(a) Both species may have the same spectrum, although this is uncommon;

(b) The substance may be decomposed by the acid or alkali: If so, an alteration in absorbance with time should be noticed. If this is found, an attempt should be made to isolate the decomposing species nearer to neutrality, or an equation like (4·4) (p. 85) can be used to obtain the absorbance of this species;

(c) The substance may have more than one ionizing group, and the two pure species isolated may differ not by one unit of charge, but by two. If this is the case, it will almost certainly be discovered during the search for the analytical wavelength (see below).

Accuracy at this stage in the determination of a pK requires that no more than 1% of one species should be present at the pH eventually chosen to record the spectrum of the other. Once the approximate pK is found, it is possible to see (from Appendix V) whether these conditions were observed. If not, the determination should be repeated. Thus, for an acid, the anionic species must be determined not less than 2 pH units above the pK_a, and the molecular species not less than 2 pH units below the pK_a.

If the unknown is a base, the procedure is exactly the same as for an acid (see above). The spectrum obtained in acid is that of the cation, and that obtained in alkali is that of the molecule.

4.7 The choice of an *analytical wavelength*

Reference to Fig. 4.1 will show that when the curves of the two pure species are plotted, one particular wavelength can be found at which the two species differ most in absorbance from one another. In Fig. 4.1, this lies at 310 nm. Once this 'analytical wavelength' is chosen, the drum, or dial, of the spectrophotometer is set at this wavelength. It is most important that the drum should not be touched until all the measurements have been taken at the various pH values. Should the wavelength drum be moved to some other value temporarily, it cannot be reset *exactly* to the same value, and a minute difference in setting during a series of determinations can cause the results to fall outside the acceptable range of variation.

The ideal analytical wavelength is that at which the one species absorbs strongly, and the other has no absorption at all. As this condition is rarely met, it is best to choose a wavelength where there is (i) a big difference in absorbance between the species, and (ii) where in both species the absorbance varies only slightly with changes in wavelength. These conditions should be sought (in decreasing order of preference) in

(a) A peak of one species over a trough in the other,
(b) A peak over a peak, provided that there is a difference of 0·2 in the absorption readings,
(c) A peak over a shoulder or inflection,
(d) A trough over a trough.

At this stage, fresh solutions should be made up from stock, and the absorbance of both species redetermined at the analytical wavelength. This gives the A_I and A_M values for use in the final calculations. This is the stage at which tomake sure that 10^{-4}M is actually the most suitable dilution i.e. one that places both A_I and A_M values in the most sensitive absorbance measuring range of the instrument. The same criterion governs the choice of 1-cm or 4-cm cells (the former is usually preferable).

4.8 Preliminary search for an approximate value of pK_a

The stock solution is diluted as before but into a buffer of such a pH that the unknown substance is only partly ionized. It is here that some knowledge of the ionization constants of common groups, and the effect on these constants of further substituents in the molecule, is valuable. This information can be obtained from Chapter 9. The absorbance of this dilution is measured, and the pK calculated from the appropriate equation (4.2 or 4.3).

4.9 Exact determination of pK_a

Using this rough estimate of pK, seven buffer solutions are made at pH values numerically equal to this pK + 0, 0·2, 0·4, 0·6, −0·2, −0·4 and −0·6, respectively. A set of seven values of pK are then obtained from measurements of the spectra of these solutions. Because of instrumental uncertainty, each absorbance reading should be taken at least three times after a random turn of the absorbance-measuring control.

The seven values are averaged as on p. 10. The spread should lie within ±0·06 unit. If the spread is wider, the entire determination should be repeated (however, a higher spread, e.g. ±0·1, is permissible when working at values below pK 0).

With practice, the whole series of operations described above can be accomplished in one working day. If a recording spectrophotometer is used for the preliminary measurements, this time can be shortened. A valuable feature of this type of instrument is the family of graphs produced, all of which should intersect at one (or more) point. Failure to show such an *isosbestic point* indicates presence of a *medium effect* which can be overcome by a graphic procedure (Johnson *et al.*, 1965).

4.10 Worked examples

Acridine has been selected as an example of a base the pK_a of which is appropriately obtained by spectrometry. The aqueous solubility at 20°C is only 0·0003M, which is too dilute for potentiometric titration without special precautions (as on p. 42).

The stock solution of acridine was prepared as described in Table 4.3. To find a suitable analytical wavelength, 5 ml stock solution was added to 5·0 ml 0·1N HCl and diluted to 50 ml with water. This solution was 0·0002M with respect

Table 4.3 Determination of the ionization constant of a mono-acidic base

Substance: Acridine, $C_{13}H_9N = 179·21$. *Temperature:* 20°C.
Concentration: 0·0002M (2×10^{-4}M; M/5000). Recrystallized acridine, m.p. 110–111°C, was dried overnight ($CaCl_2$, 20 mm, 20°C), and 0·0896 g was dissolved in 10 ml 0·1N HCl and diluted to 250 ml with glass-distilled water. This stock solution (0·002M in acridine and 0·004M in HCl) was diluted ten-fold as indicated in the text.
Analytical wavelength: 403 nm. *Cells:* 1 cm.
Absorption of neutral molecule (A_M) = 0·025.
Absorption of cation (A_I) = 0·608.

1	2	3	4	5	6
pH	A	$A_I - A$	$A - A_M$	$\log \dfrac{A - A_M}{A_I - A}$	$pK_a (= pH + \text{column 5})$
6·30	0·125	0·483	0·100	− 0·68	5·62
6·10	0·170	0·438	0·145	− 0·48	5·62
5·89	0·235	0·373	0·210	− 0·25	5·64
5·68	0·299	0·309	0·274	− 0·05	5·63
5·47	0·367	0·241	0·342	+ 0·15	5·62
5·27	0·429	0·179	0·404	+ 0·36	5·63
5·08	0·474	0·134	0·449	+ 0·53	5·61
4·85	0·523	0·085	0·498	+ 0·77	5·62

Result: $pK_a = 5·62 \pm 0·02$, at 20°C and $I = 0·01$ (using all eight values in the set).

to acridine, in 0·01N HCl (pH 2). The absorbance of this solution was measured (in 1-cm cells) over a series of wavelengths using 0·01N HCl in the blank cell. A maximum ($A = 0.608$) was found at 403 nm. The same result was obtained in 0·1N HCl (pH 1), hence it was concluded that this peak represents the cation. The neutral molecule was found to have only a low absorption at this wavelength ($A = 0.025$ both at pH 9·1 and pH 13), and its λ_{max} is at a lower wavelength. Hence 403 nm was selected for the analytical wavelength.

Next, a sighting reading was obtained by diluting 5 ml stock solution to 50 ml with phosphate–acetate buffer (0·01M previously adjusted to pH 5·5). The measured pH of the solution was found to be 5·47 and the absorbance was 0·367 (at 403 nm). Hence $pK_a = 5.62$ (from the formula in column 6 of Table 4.3). From this approximate value, the following steps led to a set of eight values. Eight solutions, all 0·0002M in acridine, were prepared by ten-fold dilution of the stock solution with buffers to give a series of pH values spaced fairly evenly through a range of 1·4 pH units, namely 0·7 unit above and below the pH numerically equal to the approximate pK_a. The pH and absorbance of these solutions were then measured giving the set of values shown in Table 4.3. These values were then averaged (as on p. 10) to give the required result.

Table 4.4   Determination of the ionization constant of a weak monoacidic base

Substance: p-Nitroaniline. $C_6H_6N_2O_2 = 138.12$.   *Temperature:* 20°C.
Concentration: 0·0001M. Recrystallized material was dried overnight ($CaCl_2$, 20 mm, 20°C) and 0·0691 g was dissolved in 500 ml glass-distilled water to give a 10^{-3}M stock solution, which was diluted ten-fold with the solutions of hydrochloric acid given in column 1, below.
Analytical wavelength: 270 nm.   *Cells:* 1 cm.
Absorption of neutral molecule (A_M) = 0·138 (pH 4·7).
Absorption of cation (A_I) = 0·693 (pH − 1·68).

1	1a	2	3	4	4a	5	6
N HCl	pH (calculated on 90% of column 1)	A	$A_I - A$	$A - A_M$	$\dfrac{A - A_M}{A_I - A}$	log of column 5	pK_a (= pH + column 5)
0·0235	1·67	0·233	0·460	0·095	95/460	− 0·68	0·99
0·0373	1·47	0·276	0·417	0·138	138/417	− 0·48	0·99
0·0607	1·26	0·335	0·358	0·197	197/358	− 0·26	1·00
0·0969	1·06	0·400	0·293	0·262	262/293	− 0·05	1·01
0·1485	0·87	0·462	0·231	0·324	324/231	+ 0·15	1·02
0·234	0·67	0·522	0·171	0·384	384/171	+ 0·35	1·02
0·373	0·47	0·570	0·123	0·432	432/123	+ 0·55	1·02

Result: pK_a (thermodynamic) = 1·01 ± 0·02 at 20°C (using all seven values in the set).

Another spectrometric determination of the ionization constant of a base is given in Table 4.4. The substance, *p*-nitroaniline, has too low a pK_a for potentiometric titration (it is 1·01), but is very conveniently handled by this method.

The spectrometric determination of the ionization constant of an acid, carried out similarly, is exemplified in Table 4.5. The example chosen (8-hydroxyquinoline) is soluble enough, and a strong enough acid, to be determined potentiometrically with considerable saving of time. However, it is interesting to compare results obtained by the two methods, and this is done in the last line of Table 4.5.

8-Hydroxyquinoline is an amphoteric substance, and there is a basic pK_a at 5·13. However, the two pKs are well spaced (see p. 53) and do not interfere with the use of either method.

Table 4.5 Determination of the ionization constant of a mono-basic acid

Substance: 8-Hydroxyquinoline. $C_9H_7NO = 145·15$. *Temperature:* 20°C.
Concentration: 0·000 05M. Recrystallized material was dried overnight ($CaCl_2$, 20 mm, 20°C), and 7.26 mg was dissolved in 500 ml glass-distilled water to give a 0·0001M stock solution. This was diluted two-fold with various buffers.
Analytical wavelength: 355 nm (the neutral molecule has λ_{max} 300 nm). *Cells:* 4 cm.
Absorption of neutral molecule $(A) = 0·045$ (pH 7·5).
Absorption of anion $(A_1) = 0·558$ (pH 13).

1	2	3	4	5	6
pH	A	$A_1 - A$	$A - A_M$	$\log \dfrac{A_1 - A}{A - A_M}$	$pK_a (= pH +$ column 5)
9·12	0·123	435	78	+ 0·75	9·87
9·32	0·167	391	122	+ 0·51	9·83
9·52	0·216	342	171	+ 0·30	9·82
9·65	0·243	315	198	+ 0·20	9·85
9·89	0·310	248	265	− 0·03	9·86
10·12	0·370	188	325	− 0·24	9·88
10·28	0·415	143	370	− 0·41	9·87
10·53	0·465	93	420	− 0·65	9·88

Result: $pK_a = 9·86 \pm 0·04$, at 20°C and $I = 0·01$ (using all eight values in the set) (cf. $pK_a = 9·89 \pm 0·03$ determined potentiometrically at 0·005M; Albert, 1953).

4.11 Activity corrections

Because the unknown is present at such great dilution, it might be assumed that the pK_a values obtained by the spectrometric method are thermodynamic. Because of the presence of buffer salts, this is not so. Thus the results are 'mixed pK_a values' as defined on p. 47, the concentration to be used in any activity correction is that of the buffer salts. For this reason it is best to use only uni-univalent buffers (see Table 4.1), in the form of sodium salts for buffers derived from acids, and as hydrochlorides for those derived from bases. The required pH is obtained by careful addition of strong acid (e.g. 1M hydrochloric acid) for the former, or by the addition of strong base (e.g. 1M sodium hydroxide) for the latter. This has the effect of maintaining the ionic strength at a constant level (0·01M) so that conversion to thermodynamic values can be made, as described on p. 50 using this value of the ionic strength.

When acidity function (H_0) solutions are used to obtain the pK_a of a *mono-acidic base*, the results are thermodynamic pK_a values. This is also the case when known *concentrations* of hydrogen ions are used (e.g. the use of 0·02N hydrochloric acid to give $p[H^+]$ 1·70, without reference to potentiometry, see Appendix III). The reason is that:

$$K_a^T = \frac{[H^+][B]}{[BH^+]} \cdot \frac{y_{H^+} \cdot y_B}{y_{BH^+}}$$

In so far as y_B is taken as 1, and y_{H^+} as equalling y_{BH^+} (see ionic activity coefficients, p. 49), it follows that $K_a^T = K_a^C$. This fortunate dispensation does not apply to acids, where y_{H^+} and y_{A^-} have to be multiplied.

4.12 Extensions of the spectrometric method

The spectrometric method can be modified for some difficult cases. Two of these are discussed below. The first pertains to the case in which the spectrum of the ionized species could not be measured and the second gives a method for measuring the pK values of two groups which, in part, ionize simultaneously.

(a) *The pK$_a$ of a very weak acid (graphical treatment)*

Sometimes a particular ionic species cannot be isolated because it decomposes, or because the solution would be too alkaline for accurate work. In such cases (the last named often arises with acids of $pK_a > 11$), equations (4.1) or (4.1a) can be rearranged to yield straight-line relationships which lead to the desired pK_a.

Table 4.6 illustrates this method by the determination of the pK_a of the phenolic group in salicylic acid (pK_a 13·82). The lower (carboxylic) pK_a is 3·00. The spectrum of the mono-anion was determined (at pH 9·2), and other spectra in 0·681N NaOH and in 1·226N NaOH. All three spectra proved to be different and it was inexpedient to use stronger alkali for several reasons, of which the principal reason is the high absorption of light by the hydroxyl ion. Thus it was

Table 4.6 Example of the graphical treatment of spectrometric data to find the ionization constant of a very weak acid

Substance: Salicylic acid (phenolic group. *Temperature:* 20°C.
Concentration: 0·001M. Recrystallized material (AR) was dried over-night (CaCl$_2$, 20 mm, 20°C), and a stock solution in water (0.005M) was diluted five-fold in the presence of sodium hydroxide and sodium chloride, as in text, this page.
Analytical wavelength: 330 nm.
Absorption of mono-anion = 0·123 at 330 nm (pH 9·2). *Cells:* 1 cm.

1	2	3	4	5	6	7
[NaOH] (by titration)	[H$^+$]	pH	$A - A_{A^-}$	$[H^+](A - A_{A^-})$	A	pK_a^c
1·226	0·555 × 10^{-14}	14·26	1·013	5·622 × 10^{-15}	1·136	13·85
0·681	0·999	14·00	0·857	8·561	0·980	13·81
0·347	1·965	13·71	0·633	12·438	0·756	13·80
0·135	5·059	13·30	0·337	17·049	0·460	13·80
0·068	10·012	13·00	0·176	17·621	0·299	13·85

Intercept, when $[H^*](A - A_{A^-}) = 0$, is 1·535, $= A_{A^{2-}}$.
Result: pK_a^c = 13·82 ± 0·03 at an ionic strength of 1·226.

evident that a pure spectrum of the di-anion could not be isolated for use in the usual calculations. Hence a 'graphical' treatment was adopted.

The analytical wavelength (330 nm) was chosen in the usual way (p. 79). Solutions (0·001M) of salicylic acid were prepared in the molarities of sodium hydroxide (carbonate-free) shown in column 1. The ionic strength of the solutions was equalized by adding sodium chloride to four of them so that $I = 1·226$. The absorption of these solutions are given in column 6. Because $A_{A^{2-}}$, the absorption of the di-anion, is not directly obtainable, equation (4.2a) was rearranged as follows:

$$K_a^c = [H^+]\left(\frac{A - A_{A^-}}{A_{A^{2-}} - A}\right),$$

whence
$$A = A_{A^{2-}} - \frac{[H^+](A - A_{A^-})}{K_a^c}. \tag{4.4}$$

The hydrogen ion concentration $[H^+]$ was calculated from $K_w/[NaOH]$. Finally the method of least squares (see p. 56) was applied to the values of A and of $[H^+](A - A_{A^-})$. This operation gave 1·535 as the absorption of the di-anion. Using this mean value of the intercept, the values of the 'concentration pK_a' at an ionic strength of 1·226 were calculated.

(b) *Overlapping pK*$_a$ *values*

When a substance has two ionizing groups that lie within 3 pK units of one another, interference is experienced in the determination of the pK values. How this difficulty arises, and how it is overcome in the potentiometric method, was described on p. 56. In the spectrometric method, the situation is more complex because not only are the two constants unknown, but the absorbance of one of the species (i.e. the monoprotonated species) cannot be determined experimentally and must also be calculated. The two ionization processes can be represented as follows:

$$diprotonated \overset{K_1}{\rightleftharpoons} monoprotonated + H^+ \overset{K_2}{\rightleftharpoons} nonprotonated + H^+.$$

This nomenclature describes all three possible cases: diacidic bases, dibasic acids, and ampholytes. The observed absorption coefficient, ε at the analytical wavelength for a given pH value, when all three species are present, is

$$\varepsilon = \varepsilon_N \cdot F_N + \varepsilon_M \cdot F_M + \varepsilon_D \cdot F_D \tag{4.5}$$

where ε_D, ε_M and ε_N are the molar absorption coefficients for the *di*protonated, *mono*protonated and *non*protonated species, respectively. The fractions of these species present (F_D, F_M and F_N) are given by

$$F_D = \frac{[H^+]^2}{G}; \quad F_M = \frac{K_1[H^+]}{G}; \quad F_N = \frac{K_1 K_2}{G},$$

where G, the denominator, is $[H^+]^2 + K_1[H^+] + K_1 K_2$. Substituting the relevant fractions in equation (4.5), multiplying throughout by the denominator and rearranging, yields

$$[H^+]^2(\varepsilon - \varepsilon_D) + K_1[H^+](\varepsilon - \varepsilon_M) + K_1 K_2(\varepsilon - \varepsilon_N) = 0, \tag{4.6}$$

which can be written as

$$\frac{[H^+]^2}{K_1} \cdot \left(\frac{\varepsilon - \varepsilon_D}{\varepsilon - \varepsilon_N}\right) + K_2 = -[H^+] \cdot \left(\frac{\varepsilon - \varepsilon_M}{\varepsilon - \varepsilon_N}\right). \tag{4.7}$$

Equation (4.6) can be solved if one analytical wavelength is found where the *non*protonated and the *di*protonated species have similar absorbances which differ markedly from the absorbance of the *mono*protonated species. The method of Bryson and Matthews (1961), well suited for these conditions, depends upon the graphical selection of two pH values yielding equal values of ε. One of these pH values should be in the region where the ionization described by K_1 predominates, and the other pH value should be relevant to the ionization of the weaker group. Under this specific condition, equation (4.6) is developed into two simultaneous equations which can be solved graphically (or by least squares) for K_1 or K_2; the intercept yields a value for the molar absorbance of the monoprotonated species in either case.

This method cannot be applied, however, when $\varepsilon_N > \varepsilon_M > \varepsilon_D$ or vice versa.

This, in our experience, is the more general case and it requires the following treatment, which begins with successive approximations. Initially we treat the ionizations as if they were independent. Thus the first step is to divide the data into two segments; those readings, ε_1, which cover the pH range where the stronger group is ionizing and those which apply to the second process, ε_2. For the stronger group the observed molar absorption coefficient, ε_1, at the given pH values is

$$\varepsilon_1 = \varepsilon_N \cdot F_N + \varepsilon_M \cdot F_M + \varepsilon_D \cdot F_D \tag{4.8}$$

where all the terms on the right-hand side are the same as in equation (4.5). However, at the lower pH values, the terms $\varepsilon_D F_D$ and $\varepsilon_M F_M$ will predominate in the total and, although $\varepsilon_N F_N$ will not be negligible, it is convenient to leave it in this form. A term G is defined as in the derivation of equation (4.6).

$$\varepsilon_1 G = \varepsilon_D [H^+]^2 + \varepsilon_M \cdot K_1 [H^+] + \varepsilon_N \cdot F_N \cdot G.$$

If the product-terms in G are evaluated, and the result divided throughout by $K_1 [H^+]$, the equation so formed can be rearranged to

$$\left(1 + \frac{K_2}{[H^+]}\right)(\varepsilon_1 - \varepsilon_N \cdot F_N) = \varepsilon_M + \frac{[H^+]}{K_1}(\varepsilon_N - \varepsilon_1 + \varepsilon_N \cdot F_N). \tag{4.9}$$

Similar treatment for the data covering the ionization of the weaker group, for which $\varepsilon_D \cdot F_D$ is small but not negligible, yields

$$\varepsilon_2 G = \varepsilon_M \cdot K_1 [H^+] + \varepsilon_N \cdot K_1 K_2 + \varepsilon_D \cdot F_D \cdot G$$

which, when rearranged as above, becomes

$$\left(1 + \frac{[H^+]}{K_1}\right)(\varepsilon_2 - \varepsilon_D \cdot F_D) = \varepsilon_M + \frac{K_2}{[H^+]}(\varepsilon_N - \varepsilon_2 + \varepsilon_D \cdot F_D). \tag{4.10}$$

Equations (4.9) and (4.10) are used only to obtain two estimates of the unknown molar absorption coefficients of the monoprotonated species, ε_M. To initialize the process, the terms in these equations $(1 + K_2/[H^+])$, $(1 + [H^+]/K_1)$ are each placed equal to unity and the terms $\varepsilon_D \cdot F_D$ and $\varepsilon_N \cdot F_N$ equal to zero. After this initialization, the two equations are in a similar form to equation (4.4). The mean value of ε_M from the separate least squares solution of equations (4.9) and (4.10) is substituted in equation (4.7). This equation is solved also by least squares for K_1 and K_2 utilizing all the data uniformly (i.e. the data are not subdivided for this purpose). These initial estimates of K_1 and K_2 are used to calculate the correction terms $(1 + K_2/[H^+])$ and $\varepsilon_N F_N$ for further use in equation (4.9), and also to calculate the similar correction terms in equation (4.10). New values of ε_M are obtained from which the mean value is substituted in equation (4.7). The process is continued reiteratively until there is no significant difference between successive values of K_1 and K_2.

If a pH meter is used during the preparation of the buffer solutions, and under certain other circumstances, the results given by this method will be 'mixed constants'. These will introduce a small error in the calculated correction terms used in equations (4.9) and (4.10) to refine the values of the intercepts. Hence activity corrections, pertinent to the three possible types of compounds, must be made. However, unless an additional check upon the values of K_1 and K_2 is required, it is not necessary to apply activity corrections to equations (4.9) and (4.10). The corrections are applied, therefore, only to equation (4.7) which is condensed for this purpose to equation (4.11) in which X, which represents the left side (save the K terms) of equation (4.7), is linearly related to Y which represents the right side.

$$Y = K_2^M + \frac{1}{K_1^M} \cdot X, \qquad (4.11)$$

where K_1^M and K_2^M are mixed constants. As an example, we will derive the equation for the application of activity correction to a diacidic base at 25°C. The equilibria are

$$H_2B^{2+} \overset{K_1}{\rightleftharpoons} BH^+ + H^+ \quad \text{and} \quad BH^+ \overset{K_2}{\rightleftharpoons} B + H^+.$$

Thus

$$K_1^T = K_1^M \cdot \frac{y_{BH^+}}{y_{H_2B^{2+}}} \quad \text{and} \quad K_2^T = K_2^M \cdot \frac{y_B}{y_{BH^+}}.$$

To evaluate the activity coefficient terms we apply equation (3.4) which (at 25°C) becomes

$$-\log y = 0.512 \, z^2 \, I^{1/2}/(1 + 1.5 \, I^{1/2}).$$

Defining a functions (FS) of the ionic strength (I) as

$$FS = I^{1/2}/(1 + 1.5 \, I^{1/2}),$$

we obtain the following terms for the relevant activity coefficients

$$y_{BH^+} = 1/10^{0.512FS}; \quad y_{H_2B^{2+}} = 1/10^{2.048FS}; \quad y_B \approx 1.$$

Substituting these values in the equations involving the mixed constants and rearranging into a convenient form, gives

$$\frac{1}{K_1^M} = \frac{10^{1.536FS}}{K_1^T} \quad \text{and} \quad K_2^M = K_2^T/10^{0.512FS}.$$

Thus, for diacidic bases type (4.7) becomes

$$Y \cdot 10^{0.512FS} = K_2^T + \frac{1}{K_1^T} \cdot 10^{2.048FS}.$$

When used in this form, equation (4.7) yields, as intercept, the thermodynamic value for K_2 and, as slope, the reciprocal of the thermodynamic value of K_1. For convenience we define three activity functions as

$$A = 10^{0.512\mathrm{FS}}; \quad A_1 = 10^{1.536\mathrm{FS}}; \quad A_2 = 10^{2.048\mathrm{FS}}$$

which are applied to equation (4.7) as shown below for the three types of compounds.

$$\frac{Y}{A} = \frac{1}{K_1^{\mathrm{T}}} \cdot X + K_2^{\mathrm{T}} \quad \text{for ampholytes,}$$

$$Y \cdot A = \frac{1}{K_1^{\mathrm{T}}} \cdot X \cdot A_2 + K_2^{\mathrm{T}} \quad \text{for diacidic bases,}$$

and

$$\frac{Y}{A_1} = \frac{X}{K_1^{\mathrm{T}} \cdot A_2} + K_2^{\mathrm{T}} \quad \text{for dibasic acids.}$$

The calculations are best performed on a digital computer.

(c) *Computer program for overlapping values*

In the FORTRAN IV program (Table 4.9), the optical densities and the corresponding pH values are arranged in an array starting with the value for the lowest pH and continuing sequentially until all the results (N) are in memory. The number of these results to be used in equation (4.9) is K, the remaining ($N - K$) results being used in equation (4.10). The order in which the data are treated depends upon the nature of the absorbances at the analytical wavelength. Normally K results are used to obtain one estimate of ε_{M} and this process is continued reiteratively until successive values of the ionization constants, as computed using equation (4.7) for all N results, do not vary. The ($N - K$) values, used in equation (4.10) to obtain another value of ε_{M}, are then processed in a similar manner until invariant constants are obtained. The mean value of ε_{M} is then substituted in equation (4.7) and the final values of $\mathrm{p}K_1$ and $\mathrm{p}K_2$ calculated without reference to the iteration process. The case where $\varepsilon_{\mathrm{N}} > \varepsilon_{\mathrm{M}} \gg \varepsilon_{\mathrm{D}}$ is treated slightly differently in the program. Under this condition the term $\varepsilon_{\mathrm{N}} F_{\mathrm{N}}$ in equation (4.9) is large, and the error produced by the initialization of this term to zero may have the consequence of over-correcting the individual absorbance readings to give too low a value for ε_{M}. However, the analogous term $\varepsilon_{\mathrm{D}} F_{\mathrm{D}}$ in equation (4.10) will be negligible if $\varepsilon_{\mathrm{D}} \leqslant 0.05\varepsilon_{\mathrm{N}}$ and the pK values are greater than 1 unit apart. In this case, only ($N - K$) values are used to calculate one value of ε_{M} which is then used to calculate pK_1 and pK_2. The program can recognize this case, and it can also distinguish between the charge-types of compounds in order to apply the appropriate activity corrections.

The computer program which we have devised for this operation is printed at the end of this chapter (Table 4.9). Examples of its use will be found in Tables 4.7 and 4.8.

Obtaining the values of overlapping constants by this method uses essentially the same operations as were described early in this chapter except that the choice of analytical wavelength is more complex, because each compound poses its own problems. However, certain generalizations can make this task easier. For example, the spectra of *diacidic bases* may reveal that the peak of the nonprotonated species (free base) at the longer wavelength is suitable for the determination because the diprotonated species (dication) may absorb only slightly in this region whereas the monoprotonated species may show an absorbance intermediate between those of the other two species. A similar distribution of absorbances may be exhibited by *ampholytes* which are likely to show zwitterionic behaviour if their pK values overlap (see p. 126). In such a case, the longer wavelength absorption band of the anionic species (nonprotonated) may be a suitable choice. For all three classes of compound there is also the possibility that at some wavelength in the spectra, the absorbances of the diprotonated and nonprotonated species are of similar magnitude but differ from those in the intermediate pH range, tending to a maximum or a minimum value for the monoprotonated species.

Practical details

For overlapping constants we invariably use a recording double-beam spectrophotometer for the selection of the analytical wavelength and proceed as follows:

    (a) The spectra of the diprotonated and monoprotonated species are obtained and confirmed in the usual way;

    (b) The complete spectrum of the nonprotonated species is redetermined at progressively lower pH values until a change of about 5% is noticed in the absorbances at some region. Similarly the spectrum of the diprotonated species is remeasured at increased pH values until a similar change is observed. The pH values, at which the changes occurred, represent the upper and lower limits between which the major portions of the two ionizations take place. A buffer is prepared at about the mean of these pH values and the spectrum of the compound measured in it. Careful comparison of this spectrum with those of the diprotonated and nonprotonated species should reveal a likely region for an analytical wavelength and may also suggest a more suitable concentration for the determination. (If a different concentration is required, the absorbances are redetermined at the new concentration, and further scans at pH values 0·3 unit above and below the mean pH value should confirm a wavelength suitable for the determination);

    (c) A series of buffers is prepared of uniform ionic strength and should extend between the upper and lower limits of pH at 0·1 to 0·2 pH intervals. The absorbances of solutions prepared in these buffers are measured, and the absorbances of diprotonated and nonprotonated species confirmed at the analytical wavelength by using a manually operated instrument.

Table 4.7 Example of calculation, by computer program (Table 4.9), of two overlapping ionization constants using spectrometric data. 3-Aminobenzoic acid ($\varepsilon_N > \varepsilon_M < \varepsilon_D$), at 25°C.

(a) Determination of the molar absorbance for the monoprotonated species

Convergence check

K_1	K_2	EPSM(1)	EPSM(2)	EPSM(AV)	M	J
0·921 13E–03	0·123 06E–04	308·64	0·00	308·64	9	0
0·867 16E–03	0·168 71E–04	291·41	0·00	291·41	9	0
0·852 67E–03	0·181 94E–04	286·42	0·00	286·42	9	0
0·848 74E–03	0·185 61E–04	285·03	0·00	285·03	9	0
0·847 68E–03	0·186 61E–04	284·66	0·00	284·66	9	0
0·847 37E–03	0·186 89E–04	284·55	0·00	284·55	9	0
0·847 29E–03	0·186 97E–04	284·52	0·00	284·52	9	0
0·847 29E–03	0·186 97E–04	284·52	0·00	284·52	9	0
0·882 76E–03	0·154 93E–04	284·52	296·61	296·61	10	10
0·834 32E–03	0·199 37E–04	284·52	279·84	279·84	10	10
0·841 53E–03	0·192 44E–04	284·52	282·46	282·46	10	10
0·840 41E–03	0·193 50E–04	284·52	282·05	282·05	10	10
0·840 58E–03	0·193 34E–04	284·52	282·12	282·12	10	10
0·840 55E–03	0·193 37E–04	284·52	282·11	282·11	10	10
0·840 56E–03	0·193 36E–04	284·52	282·11	282·11	10	10
0·843 91E–03	0·190 17E–04	284·52	282·11	283·31	10	10

(b) Calculation of pK_1 and pK_2 of 3-aminobenzoic acid using the convergence information from (a)

Concentration: 0·8410M (E–03). *Cells:* 1 cm.

Analytical wavelength: 280 nm

Molar absorbances: $\begin{cases} \text{Diprotonated} = 681\cdot3 \\ \text{Nonprotonated} = 812\cdot1 \\ \text{Monoprotonated} = 283\cdot3 \text{ (Calc. above)} \end{cases}$

Ionic strength: 0·0160

pH	G^*	EPS†	X	Y	pK_1	pK_2
2·57	0·501	595·7	0·286 59E–05	0·345 02E–02	3·076	
2·64	0·490	582·6	0·225 69E–05	0·263 79E–02	3·065	
2·72	0·480	570·7	0·166 33E–05	0·200 32E–02	3·077	
2·77	0·472	561·2	0·138 05E–05	0·166 08E–02	3·075	
2·84	0·462	549·3	0·104 93E–05	0·129 18E–02	3·084	
2·91	0·450	535·1	0·799 01E–06	0·987 00E–03	3·083	
3·00	0·434	516·1	0·558 24E–06	0·693 99E–03	3·082	
3·10	0·413	491·1	0·373 90E–06	0·453 83E–03	3·066	
3·20	0·396	470·9	0·245 52E–06	0·306 14E–03	3·068	
4·42	0·385	457·8	0·911 91E–09	0·165 27E–04		4·811
4·48	0·398	473·2	0·673 29E–09	0·163 84E–04		4·807
4·56	0·418	497·0	0·443 70E–09	0·164 92E–04		4·797

Table 4.7(b) (*contd.*)

pH	G^*	EPS†	X	Y	pK_1	pK_2
4·66	0·440	523·2	0·261 97E−09	0·160 34E−04		4·803
4·73	0·461	548·2	0·174 93E−09	0·164 93E−04		4·788
4·80	0·480	570·7	0·115 07E−09	0·166 62E−04		4·782
4·90	0·500	594·5	0·632 24E−10	0·158 96E−04		4·801
4·98	0·521	619·5	0·351 96E−10	0·161 34E−04		4·793
5·08	0·542	644·5	0·152 10E−10	0·158 18E−04		4·801
5·19	0·566	673·0	0·249 41E−11	0·159 67E−04		4·797

Results (average): $pK_1 = 3·075$; $pK_2 = 4·798$.

 $^*G$ is defined on p. 58.

 †EPS is epsilon (ε).

Table 4.8 Determination of both ionization constants of a diacidic base, requiring separation of overlapping pK_a values

Example of calculation, by computer program, of two overlapping ionization constants using spectrometric data.

Benzidine ($\varepsilon_N > \varepsilon_M \gg \varepsilon_D$), at 20°C. (The convergence check, see Table 4.7, is not reproduced here.)

Substance: Benzidine, $C_{12}H_{12}N_2 = 184·23$

Concentration: 0·5000M (E−04) *Cells:* 1 cm.

Analytical wavelength: = 300 nm.

Molar absorbances: $\begin{cases} \text{Diprotonated} = & 0·0 \\ \text{Nonprotonated} = 16440·0 \\ \text{Monoprotonated} = & 9733·1 \text{ (calculated as in text)} \end{cases}$

Ionic strength: 0·0100

pH	G^*	EPS	X	Y	pK_1	pK_2
3·00	0·110	2 200·0	− 0·232 25E−06	− 0·585 77E−03	3·417	
3·12	0·136	2 720·0	− 0·171 50E−06	− 0·429 35E−03	3·419	
3·21	0·160	3 200·0	− 0·138 14E−06	− 0·336 89E−03	3·413	
3·30	0·181	3 620·0	− 0·106 63E−06	− 0·264 63E−03	3·428	
3·42	0·219	4 380·0	− 0·789 17E−07	− 0·186 86E−03	3·420	
3·50	0·240	4 800·0	− 0·619 91E−07	− 0·148 40E−03	3·436	
3·54	0·255	5 100·0	− 0·562 34E−07	− 0·130 47E−03	3·430	
3·64	0·290	5 800·0	− 0·430 06E−07	− 0·937 67E−04	3·426	
3·74	0·321	6 420·0	− 0·318 94E−07	− 0·666 23E−04	3·438	
3·82	0·351	7 020·0	− 0·256 64E−07	− 0·482 69E−04	3·430	
4·32	0·531	10 620·0	− 0·628 41E−08	0·807 65E−05		4·610
4·44	0·564	11 280·0	− 0·433 21E−08	0·120 53E−04		4·630
4·50	0·582	11 640·0	− 0·364 55E−08	0·139 11E−04		4·629
4·56	0·597	11 940·0	− 0·302 58E−08	0·149 57E−04		4·640
4·64	0·619	12 380·0	− 0·240 57E−08	0·165 38E−04		4·641
4·72	0·637	12 740·0	− 0·187 94E−08	0·171 47E−04		4·656
4·80	0·647	12 940·0	− 0·139 61E−08	0·160 80E−04		4·705
4·92	0·685	13 700·0	− 0·108 65E−08	0·192 74E−04		4·655
5·00	0·700	14 000·0	− 0·862 54E−09	0·193 64E−04		4·665

A typical result, obtained by this method and indicative of the computer output, is given in Table 4.7 (3-aminobenzoic acid, at 25°C), using the experimental data of Bryson and Matthews (1961). The upper part of this table calculates the absorbance of the monoprotonated species at the analytical wavelength. Table 4.8 gives an example (benzidine, at 20°C) with two overlapping *basic* constants, and is based on our own experimental data of which some was published in the first edition of this book. 3-Aminobenzoic acid has zwitterionic properties which will be discussed further in Chapter 7.

4.13 Errors, precision and accuracy

The spectrometric method, carefully performed, yields results of high precision. The set of seven values, obtained over a range of 1·5 units of pH should be rejected unless the average result falls within the range ± 0.06; with practice this can be reduced.* It is not practicable to extend the set to nine values over a range of 1·9 pH units, as is recommended in potentiometric titrations (p. 31) because the two new end-values would be too sensitive to small instrumental errors. It is emphasized, however, that a range of about 1·5 pH units should be covered because abnormalities, such as decomposition (see p. 31) or failure of one or both species to obey Beer's Law, can then be detected. The latter abnormality will negate the fundamental assumption made in the derivation of the equations used and will produce a pronounced trend in the pK_a values. We have observed this effect for some aromatic bases, usually nitro and cyano derivatives, when an analytical wavelength in the region 350–420 nm is chosen for the determination. Usually the cation does not absorb in this region whereas the molecular species absorbs quite strongly. Such a trend is observed for *p*-nitroaniline when the pK_a values are calculated from optical density measurements made at 400 nm. This trend is much diminished if 270 nm is used for the analytical wavelength, as reference to Table 4.4 will confirm. When deviation from Beer's Law is encountered, if a change of analytical wavelength does not suffice to overcome it, a concentration–absorbance calibration curve should be constructed for each species and their ratio calculated by reference to these graphs.

Provided that the other precautions commonly employed in analyses by spectrophotometry are observed, and that the spectral characteristics of the compound are favourable, results of high accuracy can be expected if a constant temperature is employed for both the pH and the optical density measurements. Jacketed cell holders which can be connected to a thermostatically controlled bath are commercially available for this purpose. If a pH meter is used in the preparation of the buffer solutions, *the precision* of the method will be determined largely by its performance. Similarly *the accuracy* of the results will depend on the care taken in determining the pH of buffer readings and, even more,

*But ± 0.1 is allowable for $pK_a < 0$.

on the quality and repeatability of the spectrophotometer's performance. The importance of taking several readings at each wavelength was emphasized in the foregoing.

A different, highly accurate spectrometric method avoids determination of the pH values of buffers at the time of each spectrometric measurement of pK_a (Bolton, Hall and Reece, 1966). This is achieved by the use of a buffer solution whose $p(a_H \cdot \gamma_{Cl})$ values*, over a range of temperature and ionic strength, are tabulated (Bates and Gary, 1961). The optical density of the compound at a suitable concentration in each buffer is measured over the range of ionic strength for which data are available. However, since the selection of a buffer suitable for the determination depends upon knowledge of the approximate pK_a value of the compound, this method is really an extension and refinement of the method described in this chapter. The reader who desires to enter further into the field of thermodynamics is referred to the original literature (see also Bolton and Hall, 1969).

4.14 Common sources of error

The commonest error is to use a spectrometric method when the much quicker potentiometric method would suffice (the limitations of potentiometry with the glass electrode are outlined on p. 25). Potentiometry, moreover, cannot fail to report a pK_a in cases where the ion and molecule have the same absorption maximum and absorbance, as happens with the cation of adenine (6-amino-purine).

Imprecision in a set of results can usually be traced to one of the following causes:

(a) The buffer solutions were used in the spectrophotometer at a different temperature from that at which their pH values were measured;
(b) The solvent in the blank cell (compensation cell) had not exactly the same composition as that used for the unknown (see p. 74);
(c) The wavelength drum was reset during the course of the measurements (see p. 79);
(d) The two spectrophotometer cells were not a matched pair.

4.15 Spectrophotometric determination of the pK_a of a substance that lacks an absorption spectrum

In the general method, outlined in this chapter, optically transparent acid–base pairs (buffers) are added in great excess to the system being investigated so as to control the pH and hence the position of the acid–base equilibrium being studied. This method can also be applied in reverse, i.e. by adding optically opaque acid–base pairs (indicators) in such small amounts that they have no

*See further Appendix VI (p. 203).

appreciable effect on the state of the system. The relative concentrations of ion and molecule in the indicator, as measured spectrometrically, reveal the pH of the system being investigated. This affords a method for the spectrometric determination of pK_a for substances lacking an absorption spectrum. The method is useful for pK_a values above 11 or below 2, and is particularly good for substances that have only limited solubility. In most other cases, potentiometric titration is equally accurate and more convenient. The working details are as follows (see also King, 1965, Ch. 5).

Nine solutions of the unknown acid are made up to contain such stoichiometric concentrations of $[HA]$ and $[A^-]$ as are given in columns 3 and 4 of Table 2.3. An indicator that changes colour in about half of these solutions is found by experimental sampling. This indicator is then added to each solution in exactly equal amounts, so as to give a final concentration of 0·000 05M (for a 0·01M solution of the unknown). Optical densities of the indicator are then measured at the λ_{max} of one species. Indicators which have only one coloured species are preferred. The ratio of the densities (α) enables the values of pK_a to be calculated from the equation:

$$pK_a = pK_{ind} + \log[HA] - \log[A^-] - \log \alpha$$

where pK_{ind} is the pK_a of the indicator. The terms $[HA]$ and $[A^-]$ need correction both for $[H^+]$ and for the displacement of buffer ratio caused by converting one species of the indicator partly into the other. If the pH of the indicator is close to that of the unknown, the latter correction is usually quite small.

Some examples of the use of indicators, for determination of pK_a, are (a) trichloroacetic acid (pK_a 0·66) by v. Halban and Brüll (1944), (b) hydroxylamine (basic pK_a 5·96) by Robinson and Bower (1961), and (c) hydrocyanic acid (pK_a 9·22) by Ang (1959).

4.16 A rapid method for the approximate measurement of pK_a

Preparative organic chemists are often eager to learn the pK of a newly isolated substance in order to devise the conditions for obtaining it in larger quantity (see Chapter 5). For this purpose, a simple, rapid, and reliable, if not necessarily highly accurate, method has been devised (Clark and Cunliffe, 1973). Essentially it consists of spectrophotometric scanning of a solution undergoing potentiometric titration. It eliminates any need to weigh the compound, make up volumetric solutions, or measure the volume of the titrant.

The equipment consists of a 100-ml beaker equipped with a magnetic stirrer, thermometer and calomel and glass electrodes connected to a pH meter. The solution contained in the vessel is circulated, via small-bore flexible tubes and a peristaltic pump, through a spectrophotometer cell and back to the beaker. An arbitrary amount of a 'universal' buffer in the titration vessel is stirred and circulated through the spectrophotometer while a little of the compound is sprinkled into the beaker until a suitable concentration for

Table 4.9   Computer Program for calculating overlapping ionization constants using spectrometric data

```
 1 C        PROGRAM SPECPK
 2 C
 3 C        CALCULATES OVERLAPPING PK VALUES FROM SPECTROSCOPIC
 4 C        DATA
 5 C
 6          COMMON PH(50),D(50),EPS(50),HACT(50),X(50),S(50),
 7         1C(50),A(50),Y(50),NSUBS(40),DENOM(50),FC(50),FM(50),
 8         2FA(50),DELTA1(50),DELTA2(50),CC(50),AA(50),SS(50),
 9         3FACT1(50),FACT2(50),XX(50),YY(50),TK1(50),TK2(50),
10         4PK1(50),PK2(50)
11 C
12 C        READ THE NAME OF THE SUBSTANCE (NSUBS), THE TOTAL
13 C        NUMBER (N) OF READINGS AND THE NUMBER (K) USED FOR
14 C        THE IONIZATION OF THE STRONGER GROUP. READ THE
15 C        CONSTANT VALUES TO BE USED. THESE ARE THE OPTICAL
16 C        DENSITIES OF THE DIPROTONATED SPECIES (DC) AND THE
17 C        NONPROTONATED SPECIES (DA) FOR THE GIVEN
18 C        CONCENTRATION (CONC) AT THE ANALYTICAL WAVELENGTH
19 C        (MWL). ALLOCATE CHARGE TYPE (KTYPE) AS FOLLOWS- ZERO
20 C        FOR AMPHOLYTE, POSITIVE INTEGER FOR DIACIDIC BASES,
21 C        NEGATIVE INTEGER FOR DIBASIC ACIDS TO CALCULATE THE
22 C        CORRECTIONS FOR THE GIVEN IONIC STRENGTH (STREN).
23 C
24     986  READ(3,10)N,K,DC,DA,CONC,KTYPE,MWL,STREN,(NSUBS(I),I=1,40)
25      10  FORMAT(2I5,2F10·3,F14·9,2I4,F10·4,/,40A2)
26          WRITE(4,20) (NSUBS(I),I=1,40)
27      20  FORMAT(1H1,////,30X,12HSUBSTANCE-   ,40A2)
28 C
29 C        READ N VALUES OF PH AND OPTICAL DENSITY (D)
30 C
31          IF (N) 999,999,99
32      99  READ (3,30)(PH(I),D(I),I=1,N)
33      30  FORMAT(2F10·3)
34          EPSC=DC/CONC
35          EPSA=DA/CONC
36          FS=STREN**0·5/(1·0+1·5*(STREN**0·5))
37          ACT=10·0**(0·509*FS)
38          ACT1=10·0**(1·527*FS)
39          ACT2=10·0**(2·036*FS)
40          DO 1 I=1,N
41          FACT1(I)=1·0
42          FACT2(I)=1·0
43          DELTA1(I)=1·0
44          DELTA2(I)=1·0
45          EPS(I)=D(I)/CONC
```

Table 4.9 (*contd.*)

```
46      1 HACT(I) = 10·0**( - PH(I))
47        WRITE (4,39)
48     39 FORMAT(//,28X,17HCONVERGENCE CHECK,//,16X,
49        12HK1,11X,2HK2,9X,7HEPSM(1),3X,7HEPSM(2),
50        23X,8HEPSM(AV),1X,1HM,3X,1HJ)
51 C
52 C      COMPUTE MOLAR ABSORBANCE OF THE MONOPROTONATED
53 C      SPECIES (EPSM) USING K RESULTS IN EQUATION (4.9)
54 C      THIS VALUE WILL BE KNOWN AS CEPT1. IF THE OPTICAL
55 C      DENSITY OF THE DIPROTONATED SPECIES (DC) IS VERY
56 C      SMALL OR ZERO, USE (N - K) RESULTS IN EQUATION
57 C      (4.10) AS DETAILED BELOW (STATEMENT 106).
58 C
59        AV = 0·0
60        CK1 = 0·0
61        CK2 = 0·0
62        CEPT1 = 0·0
63        CEPT2 = 0·0
64        M = K - 1
65        J = 0
66        CHECK = 0·05*DA
67    156 M = M + 1
68        J2 = N - K
69     11 IF(DC.LE.CHECK) GO TO 301
70        GO TO 221
71    301 IF(J2 - J) 252,21,252
72    221 IF(M - K) 21,21,252
73     21 SUMS = 0·0
74        SUMC = 0·0
75        SUMSC = 0·0
76        SUMS2 = 0·0
77        SUMC2 = 0·0
78        CK1A = CK1
79        CK2A = CK2
80        DO 2 I = 1,M
81        SS(I) = (D(I)/CONC) - DELTA1(I)
82        S(I) = SS(I)*FACT1(I)
83        CC(I) = HACT(I)*(EPSC - SS(I))
84        IF(KTYPE) 31,32,33
85     31 C(I) = CC(I)/ACT
86        GO TO 12
87     32 C(I) = CC(I)*ACT
88        GO TO 12
89     33 C(I) = CC(I)*ACT1
90     12 SUMC = SUMC + C(I)
91        SUMS = SUMS + S(I)
92        SUMSC = SUMSC + S(I)*C(I)
```

Table 4.9 (*Contd.*)

```
 93          SUMS2 = SUMS2 + S(I)**2
 94        2 SUMC2 = SUMC2 + C(I)**2
 95          FN = M
 96          DENOM1 = (FN*SUMC2 − SUMC**2)
 97          SLOPE1 = (FN*SUMSC − SUMC*SUMS)/DENOM1
 98          CEPT1 = (SUMS*SUMC2 − SUMC*SUMSC)/DENOM1
 99          CK = ABS (1·0/SLOPE1)
100          EPSM = CEPT1
101          GO TO 145
102 C
103 C        COMPUTE MOLAR ABSORBANCE (EPSM) USING EQUATION
104 C        (4.10) FOR (N − K) RESULTS. THE VALUE IS KNOWN AS CEPT2,
105 C
106      252 J = N − K
107          M1 = K + 1
108      253 SUMS = 0·0
109          SUMA = 0·0
110          SUMSA = 0·0
111          SUMS2 = 0·0
112          SUMA2 = 0·0
113          CK1A = CK1
114          CK2A = CK2
115          DO4 I = M1,N
116          SS(I) = (D(I) CONC) − DELTA2(I)
117          S(I) = SS(I)*FACT2(I)
118          AA(I) = (EPSA − SS(I))/HACT(I)
119          IF (KTYPE) 34,35,36
120       34 A(I) = AA(I)*ACT1
121          GO TO 13
122       35 A(I) = AA(I)*ACT
123          GO TO 13
124       36 A(I) = AA(I)/ACT
125       13 SUMA = SUMA + A(I)
126          SUMS = SUMS + S(I)
127          SUMSA = SUMSA + S(I)*A(I)
128          SUMS2 = SUMS2 + S(I)**2
129        4 SUMA2 = SUMA2 + A(I)**2
130          FN = J
131          DENOM2 = (FN*SUMA2 − SUMA**2)
132          SLOPE2 = (FN*SUMSA − SUMA*SUMS)/DENOM2
133          CEPT2 = (SUMS*SUMA2 − SUMA*SUMSA)/DENOM2
134          CKK = ABS(SLOPE2)
135          EPSM = CEPT2
136          J2 = 0
137 C
138 C        USE THE MEAN VALUE OF EPSM IN EQUATION (4.7)
139 C        TO CALCULATE K1 AND K2 USING ALL RESULTS (N).
140 C        THESE VALUES ARE KNOWN AS CK1 AND CK2.
```

Table 4.9 (*contd.*)

```
141  C
142    145 SUMX = 0·0
143        SUMX2 = 0·0
144        SUMXY = 0·0
145        SUMY = 0·0
146        SUMY2 = 0·0
147        DO 6 I = 1,N
148        XX(I) = HACT(I)**2*(EPS(I) – EPSC)/(EPS(I) – EPSA)
149        YY(I) = – HACT(I)*(EPS(I) – EPSM)/(EPS(I) – EPSA)
150        IF (KTYPE) 44,45,46
151     44 X(I) = XX(I)/ACT2
152        Y(I) = YY(I)/ACT1
153        GO TO 14
154     45 X(I) = XX(I)
155        Y(I) = YY(I)/ACT
156        GO TO 14
157     46 X(I) = XX(I)*ACT2
158        Y(I) = YY(I)*ACT
159     14 SUMX = SUMX + X(I)
160        SUMY = SUMY + Y(I)
161        SUMX2 = SUMX2 + X(I)**2
162        SUMY2 = SUMY2 + Y(I)**2
163      6 SUMXY = SUMXY + X(I)*Y(I)
164        FN = N
165        DENOM3 = (FN*SUMX2 – SUMX**2)
166        SLOPE = (FN*SUMXY – SUMX*SUMY)/DENOM3
167        CEPT = (SUMX2*SUMY – SUMX*SUMXY)/DENOM3
168        CK2 = ABS(CEPT)
169        CK1 = 1·0/SLOPE
170        WRITE(4,40) CK1,CK2,CEPT1,CEPT2,EPSM,M,J
171     40 FORMAT(8X,2E14·5,3F10·2,2I4)
172        IF (AV.EQ.EPSM) GO TO 121
173  C
174  C     COMPUTE CORRECTION FACTORS TO BE USED IN
175  C     EQUATIONS (4·9) AND (4·10). THESE ARE DELTA1 AND
176  C     AND FACT1, DELTA2 AND FACT2 RESPECTIVELY.
177  C
178        DO8 I = 1,N
179        FACT(I) = 1·0 + CK2/HACT(I)
180        FACT2(I) = 1·0 + HACT(I)/CK1
181        DENOM(I) = (HACT(I)**2 + CK1*HACT(I) + CK1*CK2)
182        FC(I) = HACT(I)**2/DENOM(I)
183        FA(I) = CK1*CK2/DENOM(I)
184        DELTA1(I) = EPSA*FA(I)
185      8 DELTA2(I) = EPSC*FC(I)
186  C
187  C     CHECK CONVERGENCE OF SUCCESSIVE VALUES FOR EACH
188  C     CONSTANT.
```

Table 4.9 (*contd.*)

```
189  C
190         EPSILA = ABS(CK1*10·0**(−5))
191         EPSILB = ABS(CK2*10·0**(−4))
192         DIFFA = ABS(CK1A − CK1)
193         DIFFB = ABS(CK2A − CK2)
194         IF(DIFFA.GT.EPSILA)GO TO 11
195         IF(DIFFB·GT·EPSILB)GO TO 11
196         IF(DC.LE.CHECK) GO TO 315
197         IF(M − K) 156,166,146
198     166 DO 456 I = 1,N
199         FACT1(I) = 1·0
200         FACT2(I) = 1·0
201         DELTA1(I) = 0·0
202     456 DELTA2(I) = 0·0
203         GO TO 156
204     315 AV = CEPT2
205         GO TO 147
206     146 AV = (CEPT2 + CEPT2)/2·0
207     147 EPSM = AV
208         GO TO 145
209  C
210  C      PREPARE FINAL OUTPUT
211  C
212     121 WRITE(4,41)(NSUBS(I),I = 1,40)
213      41 FORMAT(1H1,///,30X,40A2,//,32X,7HRESULTS,//)
214         WRITE(4,111) EPSC,CONC,EPSA,MWL,EPSM,STREN
215     111 FORMAT(/,10X,31HMOLAR ABSORBANCES-DIPROTONATED = F8·1,
216         112X,14HCONCENTRATION = ,E11·4,/,27X,
217         214HNONPROTONATED = ,F8·1,15X,11HWAVELENGTH = ,14,2HMU,
218         3/,26X,15HMONOPROTONATED = ,F8·1,12H(CALC ABOVE),/,61X,
219         415HIONIC STRENGTH = ,F6·4,//,11X,2HPH,4X,1HD,7X,3HEPS,
220         5 9X,1HX,12X,1HY,11X,3HPK1,4X,3HPK2)
221  C
222  C      CALCULATE PK1 AND PK2.
223  C
224         SUM = 0·0
225         DO18 I = 1,K
226         TK1(I) = X(I)/(Y(I) − CEPT)
227         PK1(I) = ALOG10(1·0/TK1(I))
228      18 SUM = SUM + PK1(I)
229         FN = K
230         AV1 = SUM/FN
231         WRITE(4,112)(PH(I),D(I),EPS(I),X(I),Y(I),PK1(I),I = 1,K)
232     112 FORMAT(10X,F4·2,F7·3,F9·1,2E14·5,F9·3)
233         SUM = 0·0
234         K1 = K + 1
```

Table 4.9 (*contd.*)

```
235          DO 19 I = K1,N
236          TK2(I) = Y(I) – SLOPE*X(I)
237          PK2(I) = ALOG10 (1·0/TK2(I))
238       19 SUM = SUM + PK2(I)
239          FN = N – K
240          AV2 = SUM/FN
241          WRITE(4,113)(PH(I),D(I),EPS(I),X(I),Y(I),PK2(I),I = K1,N)
242      113 FORMAT(10X,F4·2,F7·3,F9·1,2E14·5,9X,F7·3)
243          WRITE (4,132) AV1,AV2
244      132 FORMAT(//,64X,12HAVERAGE PK1 = ,F5·3,/,
245          172X,4HPK2 = ,F5·3)
246          GO TO 986
247      999 CALL EXIT
248          END
```

recording the spectrum is reached. The pH of the solution is then progressively changed by adding small volumes of acid or alkali and recording the spectrum and pH after each addition. Although it is not essential that the spectrophotometer be of recording type, it is an advantage to be able to see the *whole* spectrum as well as isosbestic points. Such a chart helps, too, in chosing the best wavelengths for measurements of optical absorbance. Thus, a wavelength at which the ionic species have appreciably different absorbances is chosen, and the absorbances are read off for each pure species as well as for the mixture of species at each intermediate pH value. For acids, equations 4.2a and b are used whereas bases require equations 4.3a and b (p. 72). Successive spectra are run at intervals of about 0·5 pH unit until a change in spectrum indicates that a new species is being formed, and then at closer intervals until the change is complete. The stirrer and pump are stopped during pH readings, and the flow must be kept free from air bubbles. Occasionally substances are encountered that become absorbed by the plastic tubing of the pump which must, in such cases, be dispensed with, and transfers from beaker to cell made by hand. A refinement is to replace the beaker by a jacketted vessel and arrange a temperature control of the spectrometer cells.

Two buffer solutions are recommended. Mixture A contains potassium dihydrogen phosphate, piperazine, chloroacetic acid, formic acid, acetic acid, and tris(hydroxymethyl) aminomethane. It has an initial pH of about 1·8 which rises to 11·8 by the gradual addition of 0·5 ml 10N potassium hydroxide to 50 ml of the buffer. Mixture B contains disodium hydrogen phosphate, chloroacetic acid, succinic acid, piperazine, ethylenediamine, boric acid tris(hydroxymethyl)aminomethane, butylamine, and potassium hydroxide. It has an initial pH of about 12 which the gradual addition of 0·5 ml 10N hydrochloric acid (to 50 ml of the buffer) changes to about 1·6. The buffers are optically transparent above 250 nm. If shorter wavelengths need to be explored, the reference cell of the spectrometer must be supplied with a peristaltic pump and its contents subjected to the same additions as the experimental cell.

5 Relations between ionization and solubility. Determination of ionization constants by phase equilibria

5.1 Ionization constants in preparative work

In the preparative laboratory, maximal yields can be obtained by utilizing the ionization constant of the substance being made. When, as is often the case, the medium is water and the substance is present as a dissolved salt, the maximal yield is obtained by adjusting the pH to the value that is, at least, 2 units on the far side of the pK_a. Thus an acidic solution of *p*-toluidine may be at pH 1, but the pK_a is 5·1 and hence the solution should be adjusted to pH 7·1. If the substance is insoluble in water, this rule gives the nearest pH at which the maximal yield of precipitate can be obtained; if the substance is soluble in water, this rule indicates the nearest pH at which an immiscible solvent will extract the desired substance most efficiently. The principle involved is simply that a neutral molecule is less water-soluble than the corresponding ion, and, if the neutral molecule is too water-soluble to be precipitated, it is more easily shaken out of solution than an ion. Reference to Appendix V shows that when the pH is made equal to the pK_a the substance is still 50% ionized, but when the pK has been exceeded by 1 unit of pH, the substance is about 10% ionized, and when it has been exceeded by 2 units, it is only 1% ionized.

To stabilize the preparation at the chosen pH, it is desirable to effect the liberation with a reagent (a common acid or base) which has a pK_a near to this pH, because all reagents have the best buffering capacity at a pH equal to their pK_a. Sulphuric, acetic, and phosphoric acids (with pK values of about 2, 5 and 7, respectively)* are much used as liberators on the acid side of neutrality, and they are usefully supplemented by citric acid which has three close pK values (3·1, 4·7, and 6·4). Substances useful for the alkaline range will be found

---

*2 and 7 are *second* ionization constants for sulphuric and phosphoric acids.

in Table 4.1. When liberation is begun by reagents which have pK_a values between 4 and 10, it is more economically completed by hydrochloric acid or sodium hydroxide; this is because the first-named reagents change the pH very little after a certain amount has been added, but act as excellent buffers at the chosen pH. Hence alternatively, a *salt* of the desired reagent may be used in place of the free reagent, in which case hydrochloric acid or sodium hydroxide should be added until the desired pH is obtained.

If the pH of the preparation is not stabilized in this way, the pH will change during precipitation in such a direction that ionization is increased and the yield falls.

Amphoteric substances are best adjusted to a pH that is half-way between the acidic and the basic pK_a values, or else to a pH that lies between the two pK_a values and is at least two units away from any one of them.

When preparing a new substance, it is worthwhile obtaining a sample of it, no matter how wastefully, from a small preliminary run so that the pK_a can be determined very early in the work (for a rapid method, see p. 95). With this information much better yields can be obtained from the following batches.

The principal difficulties that may be encountered on applying this method arise from the fact that many preparations are made in highly concentrated solutions so that activity effects, and even adsorption effects, supervene. In spite of this, the method has so much to offer that it should be put into operation on every possible occasion (for examples, see Albert, 1955.)

Another useful application of the method can be made when purifying a base that contains a small proportion of a weaker base. The material is dissolved in acid and adjusted to the lowest pH at which all the impurity is non-ionized, as calculated from Appendix V. The impurity can then usually be filtered off or extracted with a solvent. By this 'pre-precipitation technique', many a base can be obtained pure with less loss of material than occurs in recrystallization. Acids, including phenols, can be similarly purified in alkaline solution.

5.2 Prediction of solubility from ionization constants

The solubility of an acid in an alkaline solution depends on two properties: the ionization constant, and the intrinsic solubility of the neutral molecule. Thus heptanoic acid readily dissolves in a buffer solution of pH 7 whereas its higher homologue stearic acid is practically insoluble, although both acids have the same pK_a. The low intrinsic solubility of stearic acid provides the explanation.

The observed solubility (S_0') of an acid at a given pH is due to two terms, the solubility of the neutral molecule (a saturated solution) and the solubility of the anion (far from saturated). Thus:

$$S_0' = [HA] + [A^-] \tag{5.1}$$

but

$$[A^-] = \frac{K_a[HA]}{[H^+]} \quad \text{from equation (1.5)}$$

hence

$$S_0' = [HA] + \frac{K_a[HA]}{[H^+]}$$
$$= [HA](1 + \text{antilog}(pH - pK_a)). \tag{5.2}$$

To express S_0' at, say, pH 4·32 we write $S_0^{4·32}$.

In equation (5.2), $[HA]$ equals the intrinsic solubility, i.e. the molar concentration of a saturated solution in which ionization has been prevented. Thus we may write:

$$S_0' = S_i(1 + \text{antilog}(pH - pK_a)). \tag{5.3}$$

The intrinsic solubility (S_i) of an acid is usually determined in 0·01N hydrochloric acid, and that of a base in 0·01N sodium hydroxide. Often the figures for S_i differ only slightly from those observed for S_0' in water, because the term $(1 + \text{antilog}(pH - pK_a))$ in equation (5.3) is negligible for those weak acids and bases that are only moderately soluble. The usefulness of equation (5.3) for acids becomes apparent in more alkaline solutions, and some examples are given in Table 5.1.

It is seen from Table 5.1 that although two homologues of acetic acid (namely lauric acid and heptanoic acid) have almost the same pK_a, they have very different intrinsic solubilities. When the pH is increased to equal the pK_a numerically, the solubilities are only doubled (these are the S_0^5 values). However, when the pH is increased to 2 units above the pK_a, the solubility of heptanoic acid is quite large (1·7M, or about 1 part in 5), whereas that of lauric acid is much

Table 5.1   Solubility of acids at various pH values (20°C)

Solubility (mol l^{-1})*	Lauric acid $C_{12}H_{24}O_2 = 200$ ($pK_a = 4·95$)	Heptanoic acid $C_7H_{14}O_2 = 130$ ($pK_a = 4·90$)	Benzoic acid $C_7H_6O_2 = 122$ ($pK_a = 4·12$)
S_i	8·5 × 10^{-6}	2·2 × 10^{-2}	2·3 × 10^{-2}
S_0^3	8·5 × 10^{-6}	2·2 × 10^{-2}	2·5 × 10^{-2}
S_0^4	9·1 × 10^{-6}	2·4 × 10^{-2}	4·0 × 10^{-2}
S_0^5	2·0 × 10^{-5}	5·1 × 10^{-2}	2·0 × 10^{-1}
S_0^6	1·1 × 10^{-4}	3·1 × 10^{-1}	1·8
S_0^7	1·1 × 10^{-3}	1·7	(18·3)

*S_i is the intrinsic solubility of the neutral molecule; S_0^3 is the observed solubility at pH 3, and so forth.

less (0·001M, or 1 part in 5000). Benzoic acid, which has about the same intrinsic solubility as heptanoic acid, but is a stronger acid, shows the expected increased solubility over heptanoic acid. Expressed in another way, the effect of bringing lauric acid to a pH that is 2 units more alkaline than the pK_a has been to ionize 99% of it; however, as the intrinsic solubility is so low, the dissolution of 99% as anion still leaves the solution completely saturated at 0·001M (total species) and no more can be dissolved at this pH.

In using equation (5.3), too much reliance must not be placed on values in excess of 0·1M because of activity effects. It must be remembered also that, whereas an ion may be supposed to have infinite solubility, the salt that it forms with the counter-ion (e.g. with Na^+) has a finite solubility which depends on the lattice energy of the solid state.

Conversely it may be required to know what intrinsic solubility a substance must possess to achieve a given concentration in, say, N sodium hydroxide. For this purpose, equation (5.3) is rearranged to:

$$S_i = S_0^{14}/1 + \text{antilog}\,(14 - pK_a). \tag{5.4}$$

If S_0^{14}, the required concentration at pH 14, is 0·1M, then:

for an acid of pK_a 5, S_i is 10^{-10}M
for an acid of pK_a 8, S_i is 10^{-7}M
for an acid of pK_a 11, S_i is 10^{-4}M.

On referring to Table 5.1, it becomes evident that even lauric acid can give a 0·1M solution in sodium hydroxide, although it barely reaches this concentration at pH 9 (S_0^9 for lauric acid is 0·09M, from equation (5.3)).

For bases, the equation equivalent to (5.3) is

$$S_0' = S_i(1 + \text{antilog}\,(pK_a - pH)) \tag{5.5}$$

5.3 Determination of ionization constants from solubilities

The solubility method for determining pK_a is attractive to remotely situated laboratories because no expensive apparatus need be used. On the other hand, it is somewhat laborious and, if ionic strength is not controlled, the results are far from thermodynamic. In spite of this, the results have usually been found to approximate to potentiometrically, or spectrometrically, determined constants within ± 0·1 unit. Analysis of the saturated solution, after equilibration, can be performed by gravimetry (after precipitation with a reagent), or colorimetry (after addition of a colour-forming reagent), or by spectrophotometry. Even when the spectra of both ion and molecule are identical (a circumstance which has often led to choice of the solubility method), this forms no barrier to use of spectrophotometry for analysing the solutions. The sensitivity of the solubility method can be greatly increased by marking the substance with a radioactive isotope.

The solubility method is applied as follows (Krebs and Speakman, 1945). First, the solubility (S_i) of the neutral molecular species is determined at a pH where it is suspected that this species will predominate. Next, two further determinations are made, 0·5 pH unit above and below that formerly used, in order to see whether the same solubility figure is obtained, a necessary condition to make sure that the original value was the true S_i. These three results should agree to within the experimental error of the analytical technique.

Next, the solubility is redetermined at a pH near to where the pK_a is suspected to be. From this result an approximate value of the pK_a is calculated from a rearrangement of equations (5.3) and (5.5) namely,

$$pK_a = pH - \log \left[(S_0'/S_i) - 1 \right] \text{ for acids} \tag{5.6}$$

and

$$pK_a = pH + \log \left[(S_0'/S_i) - 1 \right] \text{ for bases.} \tag{5.7}$$

A set of seven pK_a values is then obtained by determining solubilities at a series of pH values, preferably distributed evenly within the range $pK_a \pm 1$. The values are averaged, as on p. 10, to give the result.

Because solubilities are highly sensitive to the presence of foreign ions, it is necessary to carry out the determinations at a constant ionic strength (defined on p. 48). For the pH range 3 to 11, buffers made from substances in Table 4.1 are suitable if used at 0·005M strength in 0·1M sodium chloride.

In practice, an excess of the unknown is shaken in a thermostatic bath with these buffer solutions under an inert atmosphere, until a steady concentration of the unknown is found on analysis. The undissolved solid is then separated by centrifugation and the pH is measured at once. Three hours' shaking sufficed for the example reported in Table 5.2.

Table 5.2   Determination of an acidic ionization constant by the solubility method

Substance: Sulphadiazine. $C_{10}H_{10}N_4O_2S = 250 \cdot 3$. *Temperature:* 25°C.
$S_i = 6 \cdot 16$ mg per 100 ml, at pH 4·3.
    $= 2 \cdot 464 \times 10^{-4}$ M.

1	2	3	4	5
pH	S_0' (mg/100 ml)	$(S_0'/S_i) - 1$	log column 3	$pK_a (= pH)$ − column 4
6·01	8·5	0·38	− 0·42	6·43
6·35	11·1	0·80	− 0·10	6·45
6·82	19·4	2·15	0·33	6·49
7·23	43·5	6·06	0·78	6·45
7·56	86·0	13·00	1·11	6·45

Result: $pK_a = 6 \cdot 45 \pm 0 \cdot 04$ at I = 0·1, and 20°C, using all five values from the rather small set (Krebs and Speakman, 1945).

For amphoteric substances, S_i is conveniently found by plotting S'_0 against $1/\{H^+\}$ and extrapolating the straight line to $1/\{H^+\} = 0$.

No hydrogen- or hydroxyl-ion corrections are needed in this method because no stoichiometric concentrations are used.

5.4 Determination of ionization constants from vapour pressure, by partitioning between a pair of solvents, or by other phase equilibria

pK_a values may similarly be determined by partition methods with immiscible solvents which are shaken with a range of aqueous buffers (Farmer and Warth, 1904). Equilibration usually takes about three minutes. The pK_a of dithizone (an analytical reagent for metals) was thus obtained by partition between carbon tetrachloride and water, and found to be 4·64 (Irving and Bell, 1952; see also Irving, Rossotti and Harris, 1955).

In this method, it is usual to analyse the non-aqueous layer, either by spectrophotometry or by gas–liquid chromatography. The principal source of error occurs when a solute increases the solubility of water in the organic phase. The partition method has also been used for weak bases, e.g. diphenylsulphoxide ($pK_a - 2\cdot07$) by extracting a solution in cyclohexane with the acidity-function (H_0) solutions of sulphuric acid described in Table 4.2 (Virtanen and Korpela, 1968). A series of aliphatic ethers have been measured as bases in this way, after analysis by gas–liquid chromatography (Arnett and Wu, 1962).

The strength of diethyl ether as a base has been determined by measuring the ever-diminishing vapour pressure as the strengths of the sulphuric acid solutions (Table 4.2) in which it was dissolved were increased (Jaques and Leisten, 1964).

A preparative method has been described for separating two weak acids of similar pK_a values by countercurrent distribution between an organic solvent and an aqueous phase containing their salts (Barker and Beecham, 1960). The process uses a Craig 20-tube system modified to permit solvent flow in either direction. Fifteen double transfers served to separate *o*- and *p*-toluic acids whose pK_a values differ by only 0·47.

Paper electrophoresis has been used to determine ionization constants (Tate, 1981). Several runs were made in anionic buffers ($I = 0\cdot1 - 0\cdot5$) varying the pH from 1·5 to 12·5. Relative mobilities were measured by using glycerol and sodium *m*-nitrobenzenesulphonate as standards of zero and unit mobility, respectively. Acceptable pK_a values (averaging $\pm 0\cdot04$) were obtained for adenine, adenosine, ATP, cyclic AMP and related substances. The success of the method depends on the choice of conditions to minimize capillary flow, and on the presence of oxalate ions to overcome the retardation of anionic species by inorganic cations in the paper. No knowledge of the concentration of the specimen is needed. Repeated electrophoresis at a series of pH values is time-consuming, but not seriously so if several specimens are run simultaneously.

Suitable apparatus is that described by Markham and Smith (1951) but modified slightly to prevent evaporation from the paper. The ionization constants are calculated from the plot of relative mobility against pH, using equations modelled on those used for obtaining ionization constants by spectrophotometry, e.g. (4.2, p. 72). For the determination of just detectable amounts of scarce natural products (introducible as unpurified extracts), electrophoresis may be hard to equal.

Ion-exchange has been combined with high-performance liquid chromatography to determine pK_a values on the 20 picamole scale (de Wit, 1982).

6 Determination of ionization constants by conductimetry

The first ionization constants were determined by Ostwald very soon after Arrhenius discovered the phenomenon of ionization. Ostwald applied the law of mass action to the ionization of carboxylic acids at various dilutions. Using Kohlrausch's method to obtain the equivalent conductance (Λ_c), and Arrhenius's derivation of the degree of ionization (α) from equation (6.1), he substituted α in the mass law formula (as in equation (6.2)) and obtained constants which varied only a little with dilution.

$$\alpha = \frac{\Lambda_c}{\Lambda_0} \qquad (6.1)$$

$$K_a = \frac{\alpha^2 c}{1 - \alpha} \qquad (6.2)$$

where K_a is the acidic ionization constant, α is the degree of ionization, Λ_c is the *equivalent conductance* and Λ_0 the *limiting conductance*.

Λ_c, the equivalent conductance, is the conductance of a solution which contains 1 gram-equivalent of the solute, when placed between two electrodes which are 1 cm apart. For uni-univalent salts, Λ_c is also the molecular conductance (μ). For neither Λ_c nor μ is the size of the electrodes important. (Specific conductance is defined on p. 115.)

The equivalent conductance of salts increases on dilution to a limiting value (Λ_0) called the limiting equivalent conductance (formerly written Λ_∞, the conductance 'at infinite dilution'). This value can be found experimentally and accurately, but only for salts because the small concentration of hydrogen ions, maintained by the ionization of water, interferes with the ionization of weak acids and bases. For these, Λ_0 is obtained by summing the ionic conductances (ionic mobilities) of the individual ions, taken from tables such as those of Robinson and Stokes (1959) or Harned and Owen (1958).

The high equivalent conductances of hydrogen and hydroxyl ions, respectively 325 and 179 (at 20°C), arise from their being liberated from the ends of chains of water molecules and hence their not having to transport hydration shells as they move. Most other ions have equivalent conductances of only 40 to 70 (at 20°C).

6.1 Scope of the method

Until 1932, conductimetry was the method most used for obtaining ionization constants. In that year, co-operative work showed that potentiometry (Harned and Ehlers, 1932) could yield results just as accurate as those obtained by conductimetry (MacInnes and Shedlovsky, 1932). The test substance was acetic acid, and the results were $1·754 \times 10^{-5} (= pK_a\ 4·756)$ and $1·753 \times 10^{-5} (= pK_a\ 4·756)$, respectively, at 25°C. Thereafter the potentiometric method rapidly gained in favour as it was found to be more versatile and quicker, and to require fewer calculations. Accordingly, when a critical list of reliable ionization constants was published in 1939 (Dippy, 1939), the only conductimetric constants found acceptable were those of acids, and all of these had pK_a values in the range 1·89 to 5·15. Today conductimetry is used for determining the ionization constants of very weak acids in the pK_a range of 11 to 14 (see, p. 113). However, it is not so suitable for strong bases in this pK region, where special spectrometric and potentiometric methods (pp. 84, 87) are available for both acids and bases.

A conductimetric determination of pK_a differs particularly from the corresponding potentiometric determination in this way: the values in a set are obtained by simple dilution, and not by titration with acid or alkali at a fixed concentration. Hence each conductimetric reading may need different activity corrections and make the calculations (lengthy enough in themselves) quite tedious. Apart from this, the practical work is at least twice as time-consuming as in potentiometric titration because Λ_0 has to be obtained experimentally from a salt, as well as Λ_c from the free acid or base, before any value for pK_a can be worked out.

As ordinarily used, i.e. by successive dilutions of a solution of the unknown substance, conductimetry does not give accurate results for any acid with pK_a over 6·5. This is because of the inevitable presence of some carbon dioxide in the water used. Whereas a stronger acid can suppress the ionization of carbonic acid ($pK_a = 6·5$) an acid with pK_a above 6·5 must necessarily have its ionization partly repressed by the carbonic acid. In any case, the ionization of very weak acids is too slight to produce a sufficient concentration of hydrogen ions for accurate measurement against the background of the natural conductivity of water. Thus for an acid of pK_a 9, a 0·1M solution is needed to obtain significant readings, for an acid of pK_a 8, a 0·01M solution, and so on: if these are the initial concentrations, significant readings cannot persist through many dilutions.

A more successful treatment of weak acids is to find how much they inhibit

Table 6.1 Determination of the ionization constant of an acid by conductimetry

Substance: Acetic acid ($C_2H_4O_2 = 60 \cdot 05$). $\Lambda_0 = 390 \cdot 7$.
Temperature: 25°C.

(a) Without corrections

1	2	3	4
Conc. (mol l^{-1})	Λ_c	Λ_c/Λ_0	pK_c^a (from equations (6.1) and (6.2))
0·052 30	7·20	0·018 43	4·743
0·020 00	11·56	0·029 61	4·743
0·009 842	16·37	0·041 89	4·744
0·005 912	20·96	0·053 64	4·745
0·002 414	32·21	0·082 47	4·747
0·001 028	48·13	0·123 2	4·750
0·000 218 4	96·47	0·247 0	4·752
0·000 111 4	127·71	0·327 0	4·753
0·000 028 01	210·32	0·538 4	4·754

Result: Concentration $pK_a = 4 \cdot 75 \pm 0 \cdot 01$ at 25°C, at $I = 9 \times 10^{-4}$ to 2×10^{-5}, and concentration 5×10^{-2} to 3×10^{-5}

(b) With corrections

5	6	7	8
Λ_c (repeated from column 2)	α (corrected for mobility by equation (6.9))	pK_a (from column 6 by equation (6.2))	pK_a^T (from column 7, corrected for activity by equation (6.13))
7·20	0·018 65	4·732	4·764
11·56	0·029 87	4·735	4·760
16·37	0·042 22	4·737	4·758
20·96	0·054 01	4·739	4·757
32·21	0·082 90	4·743	4·757
48·13	0·123 8	4·745	4·757
96·47	0·247 7	4·749	4·757
127·71	0·327 7	4·750	4·756
210·32	0·539 3	4·753	4·756

Result: Mobility-corrected $pK_a = 4 \cdot 74 \pm 0 \cdot 01$
Result: Thermodynamic $pK_a = 4 \cdot 76 \pm 0$
MacInnes and Shedlovsky (1932)

the conductivity of hydroxyl ions in a solution of sodium hydroxide. For example, the ionization constant of boric acid (pK 9·25) was successfully determined from the conductivity of solutions containing various proportions of ammonia and boric acid (Lundén, 1907). The more difficult case of trifluoro-ethanol (pK 12·37) was similarly dealt with by observing the conductivity of solutions containing varying proportions of this substance and sodium hydroxide (Ballinger and Long, 1959). It will be at once recognized that these are actually conductimetric titrations, a technique which suffers from two disadvantages, (a) a small amount of any strongly acidic impurity in the unknown acid introduces a large error, and (b) although the conductivity due to the hydroxyl ions is diminished by the unknown acid, the conductivity due to the sodium ions (which have no less than one-quarter of the conductivity of hydroxyl ions) remains throughout the titration (see Table 6.2).

Disadvantages
The almost inevitable exposure of the unknown substance to carbon dioxide has prevented conductimetry from being widely used for determining the ionization constants of bases. Another disadvantage is that conductimetry is very sensitive to small changes in temperature. Conductimetry is not easily adapted to determining the second ionization constant in a substance with two ionizing groups. Because of the high conductivity of salts, conductimetry is not suitable for determining ionization constants at constant ionic strength.

   A special advantage of conductimetry is that significant results can be obtained at great dilution, provided that the acid has a sufficiently low pK_a. Thus Table 6.1 shows that an accurate value can still be obtained for acetic acid (p$K = 4·76$) at $2·8 \times 10^{-5}$M, a concentration at which potentiometry would fail (see pp. 62–64).

6.2 Apparatus

Conductance is the reciprocal of resistance. Hence the first requirement is to measure the specific resistance of the solution, i.e. the resistance of a cube (of the solution) of 1 cm side. A suitable arrangement of the platinum electrodes is shown in Fig. 6.1, and this fits neatly into a cell of borosilicate glass or silica. The electrodes (which may be bought in many other designs) must be lightly coated with finely divided platinum so that they have a greyish appearance. This coating needs renewing, as described in the literature (Shedlovsky and Shedlovsky, 1971), from time to time (perhaps monthly). As the volume contained between the electrodes varies from cell to cell, the first measurement to be made is the specific resistance of a standard solution of potassium chloride (Jones and Prendergast, 1937). This gives the cell-constant of the cell, and all future resistance measurements have to be multiplied by this figure to convert them to specific resistances. If the cell is rigidly made, it should retain the same cell-constant until it is replatinized.

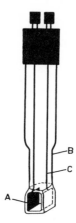

Figure 6.1 Electrode unit for conductivity cell. A, one of the two platinum electrodes; B, protective mantle of borosilicate glass or silica; C, electric connections.

A good thermostat is required for conductimetry because conductivity usually increases by 2% for each degree rise in temperature. A thyratron-regulated thermostat, constant to $0.005°C$, is recommended for fine work, and the filling liquid should be oil rather than water because of the interaction between the latter and alternating current.

A plentiful supply of conductivity water is required. This may be obtained by re-distilling distilled water containing alkaline potassium permanganate in an atmosphere free from carbon dioxide, or it can be obtained more simply from an ion-exchange column. A specific conductivity of from 1 to 10×10^{-7} ohm^{-1} cm^{-1} is suitable for work with acids of $pK_a < 6.5$.

The measuring instrument is a Wheatstone bridge of which the two ends are connected to the conductivity cell. In fine work, the connection is made indirectly to avoid transfer of heat. One arm of the bridge can be loaded with standard resistances, usually from a six decade box measuring up to 111 111 ohms. An alternating current, often 1000 cycle s^{-1}, is introduced by an oscillator and the balance point of the bridge is detected by a telephone or, preferably, a miniature cathode ray tube. Efficient earthing of the bridge is essential. A small variable condenser is included in the circuit to sharpen the readings. Detailed descriptions of various forms of electrodes and other apparatus are available (Shedlovsky and Shedlovsky, 1971).

Various compact commercial sets are offered which incorporate the Wheatstone bridge circuit, oscillator, amplifying circuit, decade resistance box and cathode tube indicator. Of these, we instance the Philips Conductivity Measuring Bridge PW 9505 on which the resistance can be directly read and has only to be multiplied by the cell-constant to give the specific resistance. This arrangement eliminates the need to solve equation (6.3).

$$R_{\text{solution}} = R_{\text{decade box}} \frac{x}{100 - x} \qquad (6.3)$$

where the slide-contact divides the bridge-wire, at the point of balance, in the ratio $x/(100 - x)$. Obach's tables for values of $x/(100 - x)$ will be found in most handbooks of chemistry.

6.3 Procedure

After the cell-constant has been measured (see p. 113), a decision must be taken whether to work in molarities (g l^{-1}) or in molalities (g kg^{-1}), both of which are about equally favoured in the literature. The attractiveness of the latter system is that weighing is independent of temperature effects, so that the bulk of the solution (and in conductivity work there is usually a great bulk of solution) does not have to be kept in the thermostat.

Any volume of solution may be used provided that the electrodes are covered to a depth of at least 1 cm. If the unknown acid is plentiful, each dilution may be made up, independently of the others, from separate weighings. Alternatively a stock solution may be made and used to fortify an initially weak solution, which thus becomes stronger progressively. The latter method should be attempted only if it is known that the electrodes are only lightly platinized (the adsorptive powers of heavily platinized electrodes can greatly reduce the concentration of a dilute solution). Finally, one can begin with a concentrated solution from which a given quantity is removed after each measurement and replaced by the same quantity of water. This sequential dilution technique, if carried too far, tends to produce cumulative errors. If the quantity is withdrawn by pipette, a new marking will be required on the stem of the pipette to indicate a 'withdrawal' instead of a 'delivery' volume.

The procedure for obtaining an ionization constant is to place a solution of the unknown substance in the conductivity cell (p. 114), and measure the specific resistance (p. 113) in Ω cm, at the required temperature. The specific resistance, when converted to its reciprocal, becomes the specific conductance in ohm^{-1} cm^{-1} and is denoted by the symbol L (the Greek letter kappa, as formerly used, caused confusion with K, the symbol for the ionization constant). The solution in the cell is then made stronger or weaker as described above, and equilibrated for temperature. The new specific resistance is measured, and each value of the specific conductivity (L) is converted to the *equivalent conductance* (Λ_c), thus

$$\Lambda_c = 1000L/c \qquad (6.4)$$

where c is the concentration in mol l^{-1}, or mol kg^{-1}, as the case may be. The equivalent conductance is suitably tabulated alongside the concentration to which it refers, as in Table 6.1.

Before any calculations can be made, values of Λ_0, the *limiting conductance*, should be obtained by a fresh series of experiments on the corresponding salt (they cannot be so quickly derived from the unknown substance itself). To this end, one equivalent of the unknown acid is dissolved in 0·98 equivalent of

Table 6.2   Limiting equivalent conductance (Λ_0) of ions in
water at various temperatures (ohm^{-1} cm^2)*

	0°	18°	25°	35°	100°
H$^+$	225	315	350	397	630
OH$^-$	105	171	199	–	450
Na$^+$	27	43	50	62	145
K$^+$	41	64	74	88	195
Li$^+$	–	–	39	–	–
NH$_4^+$	40	64	74	–	180
NEt$_4^+$	16	28	33	–	–
Cl$^-$	41	66	76	92	212
NO$_3^-$	40	62	71	–	195
ClO$_4^-$	37	59	67	–	185
Acetate	20	35	41	–	–

*These values are from various sources but agree with the values for
K$^+$, Na$^+$, Li$^+$, H$^+$, Cl$^-$ and NO$_3^-$ at 25°C of MacInnes *et al.* (1932).

0·01N sodium hydroxide (carbonate-free). The small excess of acid makes little
difference to the conductivity whereas a small excess of hydroxyl ions arising
by hydrolysis would make a big difference (see Table 6.2). The specific resistance
of this solution is measured at a range of dilutions, and the reciprocal of this
value is converted to values of Λ by equation (6.4). Each value of Λ is then
plotted as $1/\Lambda$ against $c^{1/2}$, and Λ_0 is obtained by extrapolation (Shedlovsky,
1938). From the value of Λ_{0salt}, the Λ_{0acid} is obtained by subtracting the limiting
conductance of the sodium ion and adding that of the hydrogen ion. Table 6.2
supplies these values.

Returning to the Λ_c values obtained directly on the acid, the conductance
ratios Λ_c/Λ_0 can now be worked out and added to the table (as in Table 6.1).
If this conductance ratio is taken as identical with α (the fraction ionized) from
equation (6.1), the ionization constant (K) can be calculated from equation (6.2).
The various values in a set have only to be averaged to give the required answer.
This, however, is only a *concentration* ionization constant, and the question of
applying mobility and activity corrections remains to be discussed in the next
section.

A method is also available for simultaneously obtaining highly precise values
of Λ_0 and pK_a without the extra labour of measuring the limiting conductance
of a salt (Ives and Pryor, 1955). However, the calculations are lengthier.

For further details of conductimetric apparatus, methods, and calculations,
see Shedlovsky and Shedlovsky, 1971.

In conductimetric work, a set usually consists of from 9 to 12 values. In early
studies, these values were usually obtained by a long sequence of two-fold
dilutions. It is now much more usual to work with only four dilutions, repeating

the measurements twice more on freshly prepared solutions. The dilutions may be closer than two-fold, or more widely spaced.

6.4 Refinements of calculation

Table 6.1 shows, in column 4, a series of values (for the pK_a of acetic acid) which are in excellent agreement. These were obtained with a maximal ionic strength* of 0·0009. But even 0·23M acetic acid is only 0·9% ionized, and hence the ionic strength is only 0·002. Thus it is evident that concentrations of acid which would create a high ionic strength in a potentiometric titration, give fairly low ionic strengths in conductimetric work unless the acid is a very strong one. For such substances, activity corrections make little difference to the result. The intense study of activity effects by workers in conductimetry arose from their desire to extend the method (a) to stronger acids and (b) to conductimetric titrations, both of which involve high ionic strengths; they were also pursuing activity effects to discover the nature of interionic forces.

Debye and Hückel (1923) derived, from theoretical reasoning, the expression:

$$\log K_a^T = \log K_a^C - 2A(\alpha c)^{1/2} \tag{6.5}$$

where K_a^T is the thermodynamic ionization constant, i.e. one that does not vary with the concentration, K_a^C is the concentration-dependent constant, A is a constant for the solvent at a given temperature, α is the fraction ionized, and c the concentration. This has already been partly discussed on p. 48, and comparison with equation (3.4) is relevant.

For water at 25°C, A is 0·5115, so that for uni-univalent electrolytes, where I is c,

$$\log K_a^T = \log K_a^C - 1·023(\alpha c)^{1/2}. \tag{6.6}$$

This correction has been much used for refining results at low ionic strengths, but it does not smooth results above $I = 0·01$ as much as was hoped.

A little later, Onsager (1926, 1927) made a different approach to the topic by postulating two sources A and B of interionic attractions in conductimetry, and wrote:

$$\Lambda_T = \Lambda_0 - (A + B\Lambda_0)c^{1/2}. \tag{6.7}$$

In water at 25°C, equation (6.7) becomes:

$$\Lambda_T = \Lambda_0 - (60·64 + 0·2299\Lambda_0)c^{1/2}. \tag{6.8}$$

(It should be noted that Onsager's A is different from Debye and Hückel's A.)

The next advance was made by MacInnes and Shedlovsky (1932), who applied both an Onsager-type dynamic correction for varying ionic mobility and a Debye–Hückel-type static correction for residual interionic effects.

*Ionic strength is defined on p. 48.

These authors introduced the concept of Λ_ε to replace Λ_0. This *equivalent ionic conductance* (Λ_ε) is the equivalent conductance of the totally ionized fraction, not at infinite dilution but at the concentration at which the partly ionized acid is being measured. This use of Λ_ε (which is concentration-variable) in place of Λ_0 (which is independent of concentration) was intended to compensate for the steady decrease in ionic mobility with increasing concentration. Thus equation (6.1) becomes:

$$\alpha = \Lambda_c / \Lambda_\varepsilon. \tag{6.9}$$

Values of Λ_ε can obviously be obtained as follows:

$$\Lambda_{\varepsilon\,\text{acid}} = \Lambda_{c\,\text{HCl}}\Lambda_{c\,\text{NaCl}} + \Lambda_{\alpha c\,\text{salt of acid}} \tag{6.10}$$

whence

$$\Lambda_\varepsilon = 390\!\cdot\!59 - 148\!\cdot\!61(\alpha c)^{1/2} + 165\!\cdot\!5\,\alpha c(1 - 0\!\cdot\!2274(\alpha c)^{1/2}) \tag{6.11}$$

which is obtained from a series of such equations as:

$$\Lambda_{c\,\text{NaCl}} = 126\!\cdot\!42 - 88\!\cdot\!53c^{1/2} + 89\!\cdot\!5c(1 - 0\!\cdot\!2274c^{1/2})*. \tag{6.12}$$

To find each value of Λ_ε involves a series of approximations, the first of which is made by inserting the limiting value of Λ_ε, namely $\Lambda_{0\,\text{acid}}$ into equation (6.1). This value of α is inserted into equation (6.11). A new value of α is thus found and this, in turn, leads to new value of Λ_ε, and so on until repetition does not change the result. Usually three rounds of approximation suffice.

The values of α obtained in this way are shown in column 6 of Table 6.1, and from them values of pK_a (shown in column 7) are derived by equation (6.2). It can be seen that the values in column 7 are less precise than those, so much more simply obtained, in column 4. However, the authors considered the values of column 7 to be more accurate, and they attributed the greater precision of column 4 to a partial cancelling of two opposite errors.

These mobility-corrected ionization constants were at once submitted by MacInnes and Shedlovsky to a Debye–Hückel-type correction:

$$\log K_a^{\mathrm{T}} = \log K_a^{\mathrm{C}} - 1\!\cdot\!013(\alpha c)^{1/2} \tag{6.13}$$

which is a minor variant of equation (6.6), differing principally in that the α of (6.13) is already corrected for mobility. The final values obtained from these calculations, shown in column 8 of Table 6.1, are seen to have become highly smoothed, and give a pK_a for acetic acid of $4\!\cdot\!76 \pm 0$.

This two-stage method of MacInnes and Shedlovsky has been widely used and is considered to be highly accurate for activity corrections up to $I = 0\!\cdot\!02$. A more complex formula, based on similar principles but requiring many successive approximations, was put forward by Fuoss and Kraus (1933). This has proved suitable for concentrations up to $0\!\cdot\!1\mathrm{M}$, and pK values (for acids)

---

*Shedlovsky (1932).

as low as -1. A slight simplification of this formula is available (Shedlovsky, 1938). It may seem, from the example of Table 6.1, that activity corrections are unnecessary to attain the degree of precision required for routine determination of pK_a values, say, ± 0.06. However, activity correction is essential (i) for weaker acids which must be measured either in very concentrated solutions, or in the presence of much sodium hydroxide (see p. 111), and (ii) for strong acids and bases, because they produce a high ionic strength.

7 Some other methods for the determination of ionization constants

7.1 Raman spectrometry

As traditionally practised, Raman technique required aqueous solutions to be so highly concentrated that activity corrections became meaningless. Introduction of the laser source, about 1970, permitted examination of more dilute solutions. Nevertheless, the method has not been much used, probably because only *small* molecules give spectra that are simple enough for meaningful analysis.

The substance to be determined is placed in a series of aqueous buffers, or H_0 solutions (Table 4.2), and the band heights, or the areas, contributed by either the protonated or the deprotonated group being studied is measured. This measurement is made relative to a CH-band in the same molecule, choosing a band that remains unchanged during the protonation: this is done as a safeguard against drift during the measurements.

Most of the compounds examined have been acids or bases, whose pK_a values lie outside the normal range for potentiometry (pK_a 1·3–11) and whose ionic species are indistinguishable in ultraviolet absorbance. The great majority, in fact, have been examined in the H_0 acidic region, and the results are subject to the uncertainties inseparable from this area (see p. 75).

The results obtained by Raman spectrometry usually agree with those obtained by other methods. Thus acetone gave pK_a $-7·2$ both by Raman (Deno and Wisotsky, 1963) and ultraviolet spectrometry (Campbell and Edward, 1960). Again, nitric acid furnished $-1·37$ by Raman (Young, Wu and Krawetz, 1957) and $-1·44$ by ^1H NMR (Hood and Reilly, 1960). Less close, but still comparable results have been obtained for methanesulphonic acid for which Clarke and Woodward (1966) found pK_a $-1·86$ compared to the $-1·20$ given by ^1H NMR (Covington and Lilley, 1967), both determinations being in water at 25°C. Again p-toluenesulphonic acid gave $-1·06$ by Raman spectrometry (Bonner and Torres, 1965) against the value $-1·34$ furnished by ^1H NMR (Dinius and Choppin, 1962).

One type of error to be guarded against is exemplified by the assignment of -2.2 for the pK_a of methanol, which was derived from measurement of the C–O band at 1021 cm^{-1} in aqueous sulphuric acid (Deno and Wisotsky, 1963). Unfortunately this band is very close to a strongly absorbing band in sulphuric acid. When the determination was repeated in 11M hydrochloric acid, where this type of interference does not occur, the pK_a -4.9 was obtained (Weston *et al.*, 1967).

7.2 Proton nuclear magnetic resonance

The NMR approach to the determination of ionization constants has proved attractive for the same kind of molecule to which Raman spectrometry has been applied, particularly the very strong acids and very weak bases. Like ultraviolet spectrophotometry, the NMR technique is a modified potentiometric titration, but it tends to take longer because the spinning of the containers must be started and stopped between every change in pH, and the necessary graphic work is rather time consuming.

Because the chemical shifts that arise as the pH is changed are small compared to the rate of proton exchange, only *one* band is observed for the $-$CH nearest the ionizing group regardless of the degree of ionization. This $-$CH signal progresses regularly, throughout the titration, from the position (measured in p.p.m. as δ) that represents the totally ionized form to the position that represents the totally non-ionized species. In its travel between these two points, this band corresponds to the weighted averaged of the two ionic species present in solution at that particular pH. At the end of the runs, the chemical shift (δ), which can be measured more accurately than the height of the bands, is plotted against the pH. Care has to be taken to see that a *sigmoid* curve is produced and that it extends over no more than 2.5 units of pH. Failure in these matters indicates an adverse medium effect or the wrong choice of a type of acidity-function solution (see p. 75). The pK_a is then calculated by the use of equations modelled on those used in spectrophotometry, namely (4.2) and (4.3), in which absorbances are replaced by shifts.

In general, protonation leads to a downfield chemical shift in the spectral lines that arise from the site of protonation. This shift follows the lowering of electron density at the basic centre when it is protonated. The accuracy obtainable is usually considered to be about ± 0.3 unit of pK (Cookson, 1974; Levy *et al.*, 1970), but it is occasionally more accurate.

As an internal standard, tetramethylammonium bromide has often been used but it has the disadvantage of increasing the ionic strength which is usually almost unacceptably high because of the relative insensitivity of most spectrometers. Thus, in general, a solution should be stronger than 0.1M to furnish well-resolved signals on a 60 MHz spectrometer. This insensitivity can be countered by using a more powerful instrument, say one of 360 MHz, or by taking an enormous number of readings with the help of a computer of averaged transients or similar device. Neither of these expedients may seem attractive when a series of compounds has to be measured routinely.

Sometimes an external standard is used by immersing it, enclosed in a sealed capillary, in the spinning solutions. Most frequently of all, an internal standard is found in another part of the molecule under investigation. This is usually another CH-group further from the ionizing group. In such cases, the *difference* in chemical shifts is plotted against pH. Further details of technique are available from the publications of early workers, e.g. Edward *et al.* (1962), Lee (1970), and Lee and Cameron (1971).

A few determinations of pK_a have been made in water, in spite of the interference given by side bands of the solvent (it would be easier, but less relevant, to use deuterium oxide). For examples see Handloser *et al.*, (1973), and Rabenstein and Anvarhusein (1982). Most determinations, however, have been made in H_0 solutions of aqueous sulphuric acid (Table 4.2).

Some examples of the use of ^1H NMR will now be given. Oxazole (7.1) was determined in this way (Fig. 7.1) and the pK_a found to be 0·8 whereas misapplication of potentiometry (see p. 35) had supplied the literature with much more basic values.

The following weak bases were similarly determined: sulphoxides (Landini *et al.*, 1969), aliphatic ketones (Lee, 1970), aliphatic and aromatic ketones, as well as benzaldehyde (Levy *et al.*, 1970), ethanol (Lee and Cameron, 1971), and some ethers (Bonvicini *et al.*, 1973). Some other examples will be found in Section 9.1 e.

A novel application of ^1H NMR has been found in measuring the acidic ionization of hydrocarbons in the pK_a 35–40 region, where proton exchange occurs measurably slowly. A magnesium derivative of the hydrocarbon was made, and then the rate at which mercury replaced magnesium was measured

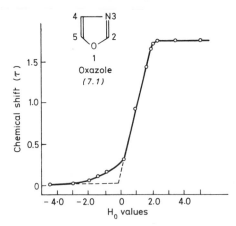

Figure 7.1 Plot of H_0 against the chemical shift of the H2 signal in fresh $DCl–D_2O$ solutions of oxazole at 33°C (Brown and Ghosh, 1969).

by observing a neighbouring $-CH$ signal:

$$R_2Mg + R_2'Hg \rightleftharpoons R_2'Mg + R_2Hg,$$

an equation that follows second-order kinetics (Dessy *et al.*, 1966). Tetramethylsilane was used as internal standard, at 33°C, and an accuracy of ± 0.3 unit of pK was claimed.

7.3 Nuclear magnetic resonance using other atoms

The principal advantage of ^{13}C over ^{1}H NMR, in determining ionization constants, is that the observed shifts are greater. Also, medium effects appear to be less (Spillane and Thomson, 1977). These authors investigated the ionization of butylsulphamic acid:

$$BuNHSO_3^- \rightleftharpoons BuNH_2^+SO_3^-$$

by four techniques which yielded the pK_a values shown in parenthesis: ^{13}C NMR (1.51 ± 0.01), ^{1}H NMR (1.57 ± 0.07), conductivity (1.64 ± 0.02), and potentiometry at 0.03M (1.92 ± 0.10). unfortunately, this potentiometric measurement was made at a concentration so low that reliable results could not be expected, see p. 35. In any case, the authors preferred the ^{13}C value, which was measured at 25 MHz, by observation of natural abundance of ^{13}C, in a spectrometer using proton decoupling and Fourier transformation. The resolution was estimated as 0.24 Hz. The substance was dissolved in 1.5M hydrochloric acid to which sodium hydroxide was added stepwise, and twelve samples examined between pH 0 and 10. Deuterium oxide was used as an internal standard.

The ionization constants of the carboxy- and mercapto-groups in the dipeptide drug captopril, which is used for lowering elevated blood-pressure, were measured in water by ^{13}C NMR, using a 50 MHz instrument (Rabenstein and Anvarhusein, 1982).

^{19}F NMR was used to find the strength, as bases, of 4-fluorobenzamide (pK_a -2.24) and 4-fluoroacetophenone (-6.20), in a mixture of acetic and sulphuric acids at 40 MHz. Tetrachlorotetrafluorocyclobutane (in carbon tetrachloride) was used as an external standard (Taft and Levine, 1962). The results agreed with those found by spectrophotometry.

2-Phenylhexafluoroisopropane, dissolved in a mixture of methanol and dimethylsulphoxide, was submitted to pK_a determination (as an acid) using ^{19}F NMR. The procedure was not to titrate with a base but to examine the effect on a $-CF$ group of exchange of the hydrogen atom in the 2-position of the isopropyl group by deuterium (also by tritium) (Klabunde and Burton, 1972).

^{31}P NMR was used to study protonation equilibrium of aliphatic and aromatic phosphine oxides (Skvortsov *et al.*, 1971).

7.4 Thermometric methods

Enthalpies of ionization were measured for 35 aliphatic and aromatic amines in a calorimeter at 25°C. First the heat of solution in carbon tetrachloride was found and this was subtracted from the heat of solution in sulphuric acid. It was noted that the ranking order of enthalpies was the same as that of established pK_a values, and this led the authors to calculate, by interpolation, a few ionization constants that were either undetermined or controversial (Arnett, Quirk and Burke, 1970).

In a related study, the heats of ionization were found, in fluorosulphuric acid, for 50 carbonyl compounds, and were converted to pK_a values by the equation:

$$-\Delta H = (1 \cdot 78 pK_a + 2 \cdot 81) \text{ kcal mol}^{-1}.$$

The values ranged from pK_a $-5 \cdot 2$ for acetophenone to $-12 \cdot 4$ for benzoyl chloride. ^{1}H NMR was used to confirm that only *one* ionic species was formed (Arnett, Quirk and Larsen, 1970).

In thermometric titrations, the unknown is placed in a Dewar flask and stirred while the titrant is added from a thermally controlled burette. The temperature of the mixture is observed with a thermistor or a multijunction thermocouple, after each of a succession of equal volumes of the titrant is added. The endpoint, if not clear, can be found by extrapolation of the branches of a graph formed by plotting temperature against the added volume. Titrations of weak acids ($pK_a < 10$) are accurate down to 10^{-3}M (Bartgel, 1976).

Although pK values can be calculated from the sigmoid curve produced in this way, the technique has been used mainly for obtaining enthalpies of ionization (e.g. Avedikian, 1966) or as a method of quantitative analysis. One of the most noteworthy uses of the method has been the determination of ionization constants of various sugars (including those in Table 9.3) by Christensen *et al.* (1970). The results, which are thermodynamic, have been obtained with great precision (e.g. $12 \cdot 61 \pm 0 \cdot 01$ at 25°C, for 2-deoxyribose) in spite of the high pH values required for the ionization of these very weak acids. For experimental details, see Christensen *et al.* (1966).

8 Zwitterions (dipolar ions)

Küster coined the work 'zwiterion' in 1897 to describe the nature of certain indicators, such as methyl orange, which act as though they have one fully ionized acidic group and one fully ionized basic group in each molecule. The names 'amphions' and 'dipolar ions' have also been put forward to describe these internal salts. But there are many amphoteric substances which are not zwitterions. Thus, a substance may have both an acidic and a basic group, but these cannot form an internal salt unless both groups are simultaneously ionized and this depends on the magnitudes of their pK_a values.

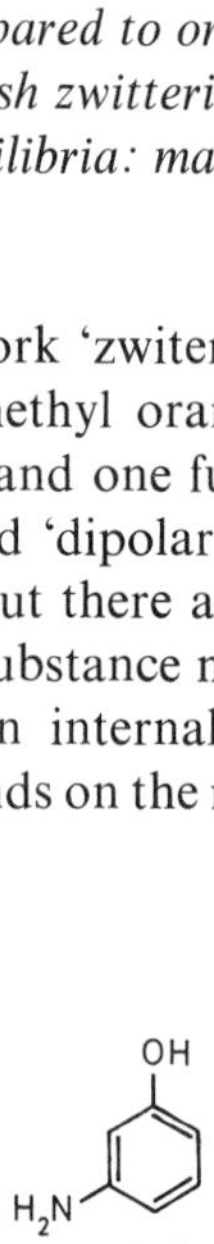

$$\overset{+}{H_3N} \cdot CH_2 \cdot CO_2H$$
(8.2)

$$\uparrow H^+$$

$$\overset{+}{H_3N} \cdot CH_2 \cdot CO_2^-$$
(8.3)

$$\uparrow H^+$$

$$H_2N \cdot CH_2 \cdot CO_2^-$$
(8.4)

(8.1)

8.1 Zwitterions compared to ordinary amphoteric substances

We will discuss *m*-aminophenol (*8.1*) as an example of an ordinary amphoteric substance, and glycine (aminoacetic acid) (*8.3*) as an example of a zwitterion.

m-Aminophenol has two pK_a values, 4·4 and 9·8 and these are similar to those of two related substances, aniline (4.9) and phenol (10·0). If the values for *m*-aminophenol are used in conjunction with Appendix V (p. 203), the following sequence of ionizations can be worked out. At, and below, pH 2·2, the basic group is completely ionized and the acidic group is not ionized; at pH 4·4 the basic group is only half ionized (all these figures are averages for a large number

126

of similar molecules considered together). Between pH 6·2 and 7·9, neither group is ionized, but at pH 9.8 the acidic group is half ionized. At, and above, pH 11·9 the acidic group is entirely ionized, and the basic group not ionized. Such substances present few problems: their pK_a values may readily be determined (as on p. 25) and assigned to the relevant ionizing groups.

Zwitterions present a different pattern of ionization. In highly acidic solutions only cations are present, e.g. (*8·2*) for glycine, and in highly alkaline solutions only anions, e.g. (*8·4*). Thus far, the resemblance to ordinary amphoteric substances is complete. The difference occurs at intermediate pH values, where the majority of molecules have *both* groups ionized. Because of the doubly charged nature of a zwitterion, it can prove confusing to speak of anion formation with alkali; and of cation formation with acids. It is better to speak of the change from (*8·3*) to (*8·2*) as 'proton gained', and from (*8·3*) to (*8·4*) as 'proton lost'.

Every molecule which has an acidic pK_a numerically lower than the basic pK_a is a zwitterion, in so far as it has acidic and basic groups strong enough to neutralize one another (Bjerrum, 1923).

Unlike the example of *m*-aminophenol, discussed above, zwitterions often have pK_a values which differ considerably from those of simpler analogues (this happens only when the acidic and basic group are physically near, or are separated mainly by a conjugated chain of double bonds). The ionization of the basic group creates a positive charge which attracts electrons and hence strengthens a nearby acidic group so that the pK_a of the latter falls. Conversely, the ionization of the acidic group creates a negative charge which releases electrons and thus makes a contribution towards strengthening the basic group (this effect would be partly neutralized by the electron-attracting (base-weakening) effect of any doubly bound oxygen atom present in the acidic group). In ordinary amphoteric substances the mutual interaction of the groups is much slighter because only one kind of group can be ionized at a time.

The pK_a values of amphoteric substances are most readily determined by potentiometry. The spectrometric method, although more time-consuming, also gives good results. The values found are, by convention, called pK_a^1, pK_a^2, pK_a^3, etc., in order of increasing pK_a. This convention is helpful at an early stage in the investigation, because it is often not possible to assign the pK_a values to a given group without much further work. If tests (see below) show that the substance is not a zwitterion, then no further work is needed, for the pK_a found during titration with acid will be that of a basic group, and the pK_a found during titration with alkali will be that of the acidic group. But when a zwitterion is titrated with acid, a proton is added to the carboxylic anion, e.g. (*8.3*) gives (*8.2*), which is quite the reverse of what happens with an ordinary amphoteric substance.

Hence the pK_a values of a new natural product may be recorded e.g. as 'pK_a (proton gained) 3·3, pK_a (protons lost) 7·4 and 9·3', and the results may remain in this state for some time until each pK_a can be assigned to its ionizing group. Thus the above figure of 3·3, obtained by titration with acid, would be that of a weak base if the substance were an ordinary amphoteric substance, but it would be that of a strong acid if the substance were a zwitterion.

8.2 How to distinguish zwitterions from ordinary ampholytes

A substance is often suspected to be a zwitterion if it is more soluble in water, less soluble in organic solvents, and has a higher melting point, than related substances with only one ionizing group. These are all signs of a salt-like character, but are insufficient as evidence of zwitterion structure because they are found in many non-electrolytes of fairly high dipole moment (say six Debye units).

Of the easily applied tests for zwitterions, the following are the best:

(a) If one of the pK_a values is markedly different from that of a partly blocked derivative, such as an ester or an O-ether, the substance is a zwitterion (see below, p. 129, for an example);

(b) If the pK_a, found by titration with acid, rises or is stationary instead of falling when the titration is repeated in 50–70% ethanol, the substance is a zwitterion (Jukes and Schmidt, 1934), (see p. 129, for an example).

However (a) and (b) could be negative, and the substance may nevertheless be a zwitterion;

(c) If the substance absorbs in the ultraviolet and one group is a carboxylic acid (the spectrum of which should not change appreciably on ionization), and the other group is an aromatic amino-group, then a large shift, in the long-wave band of the spectrum, to shorter wavelengths on adding alkali indicates that the substance is a zwitterion. This rule, slightly modified, has been used to demonstrate the zwitterionic character of the pyridine carboxylic acids (Green and Tong, 1956).

It is sometimes stated that measurement of dipole moments is a discriminating test for the detection of zwitterions, which, it is supposed, should have a dipole moment of at least 15 D (Debye units) as compared with a maximum of 6·5 D for a highly dipolar (but non-ionized) molecule such as p-nitroaniline. This was found not to be true for a series of aromatic zwitterions whose dipole moments were measured in dioxan (Serjeant, 1964). For example, p-methoxyphenyl-glycine, which was found to be over 90% zwitterionic in water, had a dipole moment in dioxan differing by only 0·18 D from that of the corresponding ester (both compounds had a dipole moment of about 2 D).

Test (b) is often negative for aromatic and heteroaromatic zwitterions which may not show the expected depression of pK_a after addition of ethanol. The test, however, has been usefully applied to aliphatic amino acids. In explanation of this, it may be recalled that ethanol, by depressing the ionization of both acids and bases, makes the pK_a of an acid higher and that of a base lower. In many aminoacids, the 'proton lost' pK_a (as revealed by titration with alkali) reacts abnormally to alcohol by not showing the expected depression of the larger of the two values. This has been attributed to the unusual nature of the equilibrium:

$$HZ^{+-} \rightleftharpoons H^+ + Z^-$$

in which all participants are charged, and the number of charges remains constant. However, amino-derivatives of sulphonic acids usually react normally to alcohol.

Glycine (8·3), when titrated in water at 20°C, shows two pK_a values, one at 2·4 (proton gained) and one at 9·8 (proton lost). The corresponding methyl ester has only one pK_a (7·6), which corresponds to the ionization of its amino-group. On numerical grounds, the value 7·6, from the ester, is more related to 9·8 than to 2·4 in glycine. Hence the value 9·8 must refer to the ionization of the amino-group of glycine, because 7·6 refers to the ionization of the amino-group in the ester. This is true, even though the 7·6 value was obtained by titrating the ester with acid, and the 9·8 value by titrating glycine with alkali (a mental barrier, which sometimes arises here, may be overcome by picturing the ester, supplied as the hydrochloride, being titrated with alkali). The interpretation, that the 9·8 value refers to the equilibrium (8·3)⇌(8·4), is confirmed by re-titration in 50% ethanol, when *both* pK_a values increase (to 2·7 and 10·0, respectively) which indicates that the substance is a zwitterion (see p. 128), and hence that the upper pK_a belongs to the ionization of the basic group.

The following factors govern the magnitude of the pK_a values of glycine. The pK_a of acetic acid is 4·8, but the ionized amino-group in the glycine zwitterion is electron-attracting, and hence acid-strengthening, and the anionic group is near enough to be strengthened by this to 2·4. The methyl ester of glycine, thanks to the inductive (− I) effect of the − CO_2CH_3 group, is 1000 times weaker a base than methylamine (the pK_a of the ester is 7·6 compared to 10·7 for methylamine). This inductive effect should be the same in the neutral species of glycine (8·5), a substance which exists in minute amounts in equilibrium with the zwitterion of glycine (8·3), as we shall discuss later. That the inductive effect would be similar in (8·5) as in its methyl ester follows from the fact that acids and their esters have similar dipole moments. That the pK_a of glycine is 9·8 instead of 7·6 is thus further proof of its zwitterion nature. The magnitude of the value 9·8 is the resultant of two opposing tendencies, (a) the base-weakening inductive effect of the carboxyl-group, and (b) the strengthening effect of an anion (through electron release) on a basic-group near enough to be influenced. Hence glycine is only eight times (0·9 pK unit) weaker than methylamine.

$$H_2N \cdot CH_2 \cdot CO_2H \qquad \overset{+}{H_3N} \cdot CH_2CH_2CH_2CH_2 \cdot CO_2^-$$
(8.5)   (8.6)

(8.7)

The effect of removing the two charged groups from one another's influence is seen in δ-aminovaleric acid ($8 \cdot 6$) which has pK_a values of $4 \cdot 3$ and $10 \cdot 8$ which are very near to those of acetic acid ($4 \cdot 8$) and methylamine ($10 \cdot 7$) respectively (the pK_a of the methyl ester is $10 \cdot 15$).

For heterocyclic zwitterions, e.g. nicotinic acid ($8 \cdot 7$) and its isomers, see Lumme (1957).

8.3 Zwitterionic equilibria: macroscopic and microscopic constants

The pK_a values determined experimentally for zwitterionic substances are often termed 'macroscopic constants' because they describe only a composite of the processes which are actually occurring, and which can be represented diagrammatically as in Scheme 8.1.

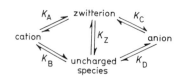

Scheme 8.1

It is obvious that the two macroscopic constants, K_1 and K_2, cannot fully describe the equilibria denoted by these four microscopic constants: K_A, K_B, K_C, K_D and the zwitterionic constant K_Z. In fact, the macroscopic constant K_1 merely reports on a gross process of which the equilibria described by K_A, K_B, and K_Z are the essential components. Likewise, the macroscopic constant K_2 is a composite of K_C, K_D, and K_Z. This situation expresses the fact that each ionizing group has *two* constants, one for when the other group is ionized and one for when the other group is not ionized.

When the two ionizing groups are arithmetically so far apart as those of glycine, where only $0 \cdot 005\%$ of neutral species is in equilibrium with $99 \cdot 995\%$ of zwitterion, knowledge of the two macroscopic constants usually suffices. However, a more detailed survey of the problem becomes necessary when the two pK values lie within 3 units of one another. For example, 4-aminobenzoic acid has a 9 to 1 ratio of neutral species to zwitterion (in water at 25°C) (van der Graaf, Hoefnagel and Wepster, 1981). This was found by use of equations devised by Adams (1916) who showed that the microscopic constants are related to the measured macroscopic constants, K_1 and K_2 by the following equations:

$$K_1 = K_A + K_B \tag{8.1}$$

$$\frac{1}{K_2} = \frac{1}{K_C} + \frac{1}{K_D} \tag{8.2}$$

$$K_Z = \frac{K_A}{K_B} = \frac{K_D}{K_C}. \tag{8.3}$$

These equations cannot be solved until one of the microscopic constants is known. As a first step in this direction, van der Graaf *et al.* (1981) set out to derive pK_A from the only pK_a of trimethylammoniobenzoic acid (3·23). This was done by replacing the Hammett sigma value of 0·55 (for N^+Me_3) by 0·38 (for N^+H_3), without changing the Bjerrum electrostatic term for the distance (0·043 nm in both cases) between the charge and the dissociating proton. These considerations gave 3·40 for pK_A, from which all the other four microconstants followed.

When a substance absorbs well in the ultraviolet, it is easiest to penetrate into the Adams equations by first finding K_Z. The basis of this method is the common experience that the spectrum of an ester (molecule) is identical with that of the neutral species of the corresponding acid (molecule, not zwitterion), whereas that of the cations of ester and acid are almost identical with that of the acid in its zwitterionic form. This leads directly to a value for K_Z which, if fed into the Adams set of equations, rapidly furnishes all the microscopic constants. No knowledge of the pK_a of the ester is required, and the following equation is used:

$$K_Z = \frac{\varepsilon_{ester} - \varepsilon_{mol}}{\varepsilon_{mol} - \varepsilon_{cation}}. \tag{8.4}$$

For this equation, ε_{mol} is derived from the measured optical absorbance (A_{mol}) at the isoelectric point (potentiometrically determined) by subtracting contributions from the anionic and cationic species present, a correction easy to apply (Bryson, Davies and Serjeant, 1963).

Much of the earlier work on zwitterionic ratios was done by Ebert (1926), based on the assumption of Wegscheider (1895) that the equilibrium constant (K_B) for adding a proton to the nitrogen atom of the aminoacid in its non-zwitterionic form is the same as that (K_E) of the methyl ester. The validity of this hypothesis was tested by Bryson *et al.* (1963) who found that a series of ten N-aryl glycines (8·8) gave the relation: $pK_B = pK_E + 0·2$. These compounds demonstrate the effectiveness of test (c) on p. 128 in that the ultraviolet spectra of the molecular species of the esters were identical with both the molecular and anionic species of the acids when the compound was non-zwitterionic (e.g. when R = *m*-nitro-).

(8.8)

(8.9)

(8.10)

When the basic strength of the amino-group, in the series (*8·8*), was increased by changing substituent R, Bryson *et al.* found an increasing divergence between the spectrum of the molecular species of the acid and that of the ester. When R was *p*-methoxy-, the spectrum of the molecular species of the acid resembled that of its cation although, fortunately, the spectrum of the anion was still identical to that of the molecular species of the ester. Hence K_Z was obtained from the latter relationship.

In all cases, the percentage in zwitterionic form can be obtained by multiplying K_Z by 100 and dividing the product by $K_Z + 1$.

It is evident that the correction (0·20) of Bryson *et al.* would make a serious difference to derivation of tautomeric ratios between, say, 1 and 100. For example, for pyridine-4-carboxylic acid, the ratio of zwitterion to molecule of 25 : 1 when derived from Ebert's method, becomes 78 : 1 when the correction is applied. However, for larger or smaller ratios, the correction makes an inconsiderable effect. It would be preferable to ignore the Ebert approach, but unfortunately many zwitterionic compounds do not possess spectra which allow the direct determination of K_Z. Generally the acidic group is much closer to the ring, and under these conditions it is unlikely that the spectrum of the anionic species will be identical to the spectrum of the molecular non-zwitterionic form of the acid.

Although these circumstances impose more difficult conditions, the microscopic constants for *m*-aminobenzoic acid have been found by obtaining estimates of the relation between pK_B and pK_E using two methods both of which decreased the basic strength of the amino-group to such an extent that the resulting compounds were predominantly non-zwitterionic (Serjeant, 1969). The first of these was the substitution of a nitro-group *meta* to both the amino- and carboxylic acid groups (*8·9*). The pK_a value of the amino group was compared with that of methyl 3-amino-5-nitrobenzoate. Because the acid was non-zwitterionic, its pK_a value was assumed to be equal to pK_B, and this allowed a direct relation to pK_E to be established. This was found to be $pK_B = pK_E + 0·08$ for the methyl benzoate derivative and $pK_B = pK_E + 0·03$ for the ethyl benzoate derivative. The second method reduced the basic strength of the amino-group in *m*-aminobenzoic acid by insertion of the group $- CH_2 \cdot COOEt$ on the nitrogen atom to form the *meta*-substituted *N*-aryl-glycinate (*8·10*). This yielded a separate estimate of the relations between pK_B and pK_E which were $pK_B = pK_E + 0·09$ for the methyl benzoate and $pK_B = pK_E + 0·04$ for the ethyl derivative.

These results were applied to *m*-aminobenzoic acid, the pK_B for the non-zwitterionic form of this acid being assumed to bear the same relation to the pK_a of its methyl ester as those given above, i.e. $pK_B = pK_E + 0·09$. The macroscopic constants for the acid given in Table 4.7 are $pK_1 = 3·08$ and $pK_2 = 4·80$ at 25°C whilst that of the ester was found by Bryson and Matthews (1961) to be 3·56. The microscopic constants can be calculated from these data as follows:

$$pK_B = 3·56 + 0·09 = 3·65$$

i.e. $K_B = 2·24 \times 10^{-4}$.

Using equation (8.1)

$$K_A = 6{\cdot}08 \times 10^{-4}$$

and hence, from equation (7.3)

$$K_Z = \frac{6{\cdot}08 \times 10^{-4}}{2{\cdot}24 \times 10^{-4}} = 2{\cdot}72.$$

Combining equations (8.2) and (8.3), and solving for K_C, yields

$$K_C = \frac{K_2(K_Z + 1)}{K_Z} = 2{\cdot}17 \times 10^{-5}$$

and hence

$$K_D = 5{\cdot}90 \times 10^{-5}.$$

Thus, the ionization processes involved can be represented as in Scheme 8.2, and K_Z (2·72), describes the process

$$NH_2{\cdot}C_6H_4{\cdot}CO_2H \rightleftharpoons NH_3^+{\cdot}C_6H_4{\cdot}CO_2^-. \qquad (8.5)$$

Scheme 8.2 $\quad NH_3^+{\cdot}C_6H_4{\cdot}CO_2H$

$pK_A = 3{\cdot}22 \nearrow \quad NH_3^+{\cdot}C_6H_4{\cdot}CO_2^- + H^+ \quad \searrow pK_C = 4{\cdot}66$

$pK_B = 3{\cdot}65 \searrow \quad NH_2{\cdot}C_6H_4{\cdot}CO_2H + H^+ \quad \nearrow pK_D = 4{\cdot}23 \quad NH_2{\cdot}C_6H_4{\cdot}CO_2^- + H^+$

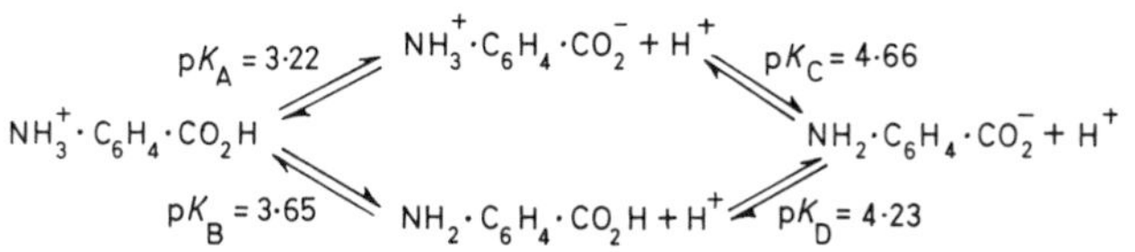

It should be noted that the equilibrium (8.5) is independent of pH. Hence the fraction present as zwitterion as a function of pH is obtained by calculating the fraction present as the monoprotonated species, using the macroscopic constants K_1 and K_2, and multiplying the result by the zwitterion fraction, F_Z. Since the latter is equal to $K_Z/(1 + K_Z)$ we obtain

$$F_Z = \frac{K_Z}{(1 + K_Z)} \cdot \frac{K_1[H^+]}{([H^+]^2 + K_1{\cdot}[H^+] + K_1{\cdot}K_2)}.$$

These two substitution method seem to be generally applicable to aromatic aminoacids whose functional groups are *meta* to one another. However, for their *para*-substituted isomers, the following complication can arise. A $-CH_2CO_2Et$ group attached to the nitrogen atom, as in (8·10), can exert a base-weakening effect so great that the pK_a value for the amino-group may be less than zero, and in this pH region, the accuracy of determination is uncertain. On the other hand, substitution of a base-weakening group in the position *ortho* to the amino-group can introduce group interactions absent from the original zwitterionic compound. In such cases it is best to use the extrapolation method introduced by Robinson and Biggs (1957) to obtain estimates of pK_B. For those approaches that depend on the introduction of substituents, a knowledge of the quantitative effect of various substituents on ionization can be obtained from a study of Hammett's *Linear Free Energy Equation* (Hammett, 1970), with its sigma and rho values, which serve this purpose admirably for aromatic compounds. For aliphatic substances, the Taft equation (Taft, 1960) can be used. A book on the prediction of pK_a values is also available (Perrin, Dempsey and Serjeant, 1981). Much that is useful for this purpose can be gleaned from Chapter 1 and the tables of Chapter 9 in the present work.

Determination of microscopic constants and zwitterionic ratios have played an important part in understanding the ionic composition of many biologically active molecules. Examples are: tyrosine (Edsall, Martin, and Hollingworth, 1958), glutathione (Martin and Edsall, 1958), tyrosine (Benesch and Benesch, 1955), and adrenaline (epinephrin) (Sinistri and Villa, 1962).

9 The ionization constants of typical acids and bases

In order to select the most appropriate method for the determination of an unknown ionization constant, the experimenter should be able to forecast an approximate value for the expected pK_a. For uncomplicated aromatic substances, this can be done from Hammett's Free Energy Equation, using its sigma and rho values (Hammett, 1940, 1970). In all cases, help can be had from a book on the prediction of pK_a for organic acids and bases (Perrin *et al.,* 1981).

Help in forecasting can also be obtained from the established pK_a values of analogous substances. Comprehensive lists of such values have been compiled for the International Union of Pure and Applied Chemistry: for *organic bases,* Perrin (1965) and large supplement (1972), for *organic acids,* Serjeant and Dempsey (1979) a listing that carries on from that of Kortüm, Vogel, and Andrussov (1961), and for *inorganic bases and acids,* Perrin (1983). In these lists, results are graded as 'reliable' when the uncertainty does not exceed $\pm 0.005 \, pK$ unit, as 'approximate' when it exceeds this figure but is less than ± 0.04 unit, and 'uncertain' when in excess of this. Such tables can also be used to suggest the structure of a new substance from its ionization constant (s) (Perrin, Dempsey and Serjeant, 1981).

The tables and text of the present chapter are intended to list the ionization constants of several hundred common substances, and to illustrate the more fundamental connections between ionization constants and structure. The following order is observed in the tables: the organic section precedes the inorganic section, organic acids precede organic bases, aliphatic substances precede aromatic substances. Saturated heterocyclic ring-systems will be found at the end of the aliphatic tables, and heteroaromatic substances at the end of the aromatic tables. Where no reference is given, the values have been taken from the IUPAC tables. Unlisted values may be sought in *Chemical Abstracts* under the name of the substance or, better, under the heading 'Ionization'.

A ORGANIC SECTION

9.1 The oxygen acids (monobasic)

(a) *Aliphatic carboxylic acids*

A good introduction to the inductive effects of substituents can be obtained by studying aliphatic carboxylic acids. Only a few substituents, e.g. the methyl-group, and the carboxylic anion, possess a positive inductive effect $(+\mathrm{I})$, and hence are acid-weakening. Most of the common substituents have a negative inductive effect $(-\mathrm{I})$, and hence are acid-strengthening. In the following lists, the groups are in approximate order of decreasing effectiveness:

$(-\mathrm{I})$ $^+\mathrm{NR}_3$, NO_2, $\mathrm{SO}_2\mathrm{R}$, CN, F, Cl (and Br), I, CF_3, COR (where R is OH, NH_2, or alkoxy), COR (where R is alkyl), OR, SR, NH_2, $\mathrm{C}_6\mathrm{H}_5$

$(+\mathrm{I})$ CO_2^-, O^-, NH^-, alkyl.

The acid-weakening effect of the methyl-group is seen by comparing acetic and formic acids in Table 9.1. Further increases in chain length have little effect, as shown (up to C_{18}) by Jukes and Schmidt (1935). The series: methylacetic (propionic), dimethylacetic, and trimethylacetic acids reveals that even the substitution of several methyl-groups into acetic acid itself causes little further decrease of acid strength. The carboxylic anion substituent is treated, later in Section 9.2 and Table 9.6.

The acid-strengthening groups are most conveniently studied as substituents of acetic acid (see Table 9.1), because not many substituted formic acids are available. Contrary to what is commonly supposed, only the first four $(-\mathrm{I})$ groups in the above list can effect a substantial change (i.e. 2 units or more) in the pK_a; but multiple insertions of a less inductive group can be very effective, as trifluoroacetic acid shows.

The acid-strengthening effect of a $(-\mathrm{I})$ group persists, although to an ever-diminishing extent, when the group is separated from the anion by one or more

Table 9.1   Aliphatic and alicyclic carboxylic acids (monobasic)

In water. For effect of temperature on pK_a, see Appendix IV. *Temperature*: 25°C, except where otherwise stated.

Acid	pK_a	°C
Homologues and isomers		
Formic	3·75	
Acetic	4·756	
Acetic-O-d	5·31[a]	
Acetic-2-d_3	4·77	
Propionic	4·87	
Valeric (C_5)	4·83	20
Octanoic	4·89	
Dimethylacetic (isobutyric)	4·84	20
Trimethylacetic (pivalic)	5·03	20
Substituted acetic acids		
Trifluoro-	0·52	
Trichloro-	0·66	20

Table 9.1 (*contd.*)

Acid	pK_a	°C
Dichloro-	1·35	
Nitro-	1·48[b]	24
Trifluoromethylsulphonyl-	1·88	
Trifluoromethylsulphinyl-	2·06	
Methylsulphonyl- ($-SO_2CH_3$)	2·36[c]	
Cyano-	2·47	
Thiocyanato- ($-SCN$)	2·58[d]	
Fluoro-	2·59	
Chloro-	2·87	
Bromo-	2·90	
Trifluoromethylthio-	2·95	
Iodo-	3·18	
Phenoxy-	3·17	
Ethoxycarbonyl- ($-CO_2Et$) (ethyl hydrogen malonate)	3·45	
Carbamoyl- ($-CONH_2$) (malonamic acid)	3·64	
Acetyl- (acetoacetic acid)	(3·6)[e]	
Ethoxy-	3·65	18
Acetamido- (acetylglycine)	3·67	
Methylthio- ($-SMe$)	3·66	
Hydroxy- (glycolic acid)	3·83[f]	
Phenyl-	4·31[g]	
Benzyl- (3-phenylpropionic acid)	4·66	
Unsaturated acids (*trans, except where marked cis*)		
Acrylic ($CH_2:CH·CO_2H$)	4·25	
Penta-2-en-1-oic	4·74	
Penta-3-en-1-oic	4·51	
Cinnamic	4·44	
Cinnamic (*cis*)	3·88	
Propiolic ($CH\vdots C·CO_2H$)	1·84	
But-2-yn-1-oic (tetrolic acid)	2·62	
Alicyclic acids		
Cyclopropane carboxylic	4·83	
Cyclopentane carboxylic	4·99	
Cyclohexane carboxylic	4·91	

[a]In D_2O.
[b]A further pK at 8·90 (nitro group)
[c]Melander (1934).
[d]Ostwald (1889).
[e]Cf. acetoformic (pyruvic) acid, 2·49, an equilibrium value arising from 1·55 (anhydrous) and 2·95 (hydrated covalently on CO group).
[f]Cf. 2-hydroxypropionic (lactic) acid 3·86
[g]Cf. Phenylformic (benzoic) acid 4·20.

methylene group(s) which are poor conductors of electrons. For example, α-trifluoroacetic, β-trifluoropropionic, and γ-trifluorobutyric acids have pK_a values of 0·23, 3·02, and 4·16, respectively. A less inductive group causes less disturbance, e.g. α-, β-, and γ-chlorobutyric acids have pK_a values of 2·84, 4·06, and 4·52, respectively.

The double-bond, and the phenyl-group, have a small ($-$ I) effect, and are also capable of a certain amount of acid-weakening through a positive mesomeric effect ($+$ M) when their double-bond can conjugate with other double-bonds. Thus phenylacetic acid is a little stronger than acetic acid, but phenylformic acid (benzoic acid) is a little weaker than formic acid, and cinnamic acid is a little weaker than acrylic acid. Any group which has mesomeric (M) properties (and only a few have not) can transmit these through one or more double bonds to the carboxylic group. The following list of mesomeric substituents may prove useful (it is important to note that, in the absence of a double-bond capable of conjugating, mesomeric effects cannot operate).

($-$ M) NO_2, CN, COR, SO_2R (weak)
($+$ M)   Halogens (fluorine is strongest), OMe, OH, NR_2 ; then (more weakly) NH, COR, O$^-$, alkyl, aryl.

Perhaps the only common substituents that lack an M effect are ions like —SO_3^- and —NH_3^+.

The heightened strength of *cis-* (compared to *trans-*) cinnamic acid has been attributed to the shorter distance separating the phenyl- and carboxylic-groups in the *cis*-isomer, permitting a ($-$ I) effect across the intervening space, a 'field effect' (Swain and Lupton, 1968).

The alicyclic carboxylic acids are similar in strength to the aliphatic acids, regardless of the ring-size. The acid-strengthening nature of cationic groups (e.g. —NH_3^+) will be met again in Section 9.8 and in Table 9.12.

(b) *Aromatic carboxylic acids*

A selection of aromatic carboxylic acids is given in Table 9.2. The values are arranged, mainly, in order of increasing acidic strength of the *meta*-substituted isomers. Each of the three positions of substitution in the benzene ring has a clear individuality. The *meta* position is the least complicated, because groups placed there exercise an almost pure ($+$ I) or ($-$ I) effect, as in the saturated aliphatic acids. Although these groups retain their sign, the extent of their inductive effect is much less, because the distance is greater, than when they are substituted in the α-position to an aliphatic carboxylic group.

In the *para* position, the (I) effect should be smaller because of increased distance. However, the *para* position is (from considerations of valency) capable of facilitating mesomerism and hence almost every substituent placed in this position is capable of exerting an (M) effect. If the sign of both (I) and (M) effects is the same, the effect is increased. Thus the methyl-group has ($+$ I) and ($+$ M) properties, and hence the strength of benzoic acid is decreased by

Table 9.2   Aromatic carboxylic acids (monobasic)

Arranged, as far as possible, in order of increasing strength of meta-substituted isomer. In water.

Temperature: 25°C, except where otherwise stated

Acid	pK_a		
Benzoic acid	4·199[a],	4·205[b],	4·204[c]
Benzoic acids	*o-*	*m-*	*p-*
Methyl-	3·91	4·25	4·37
Methoxy-	4·08	4·10	4·50
t-Butyl-	3·54	4·20	4·38
Trimethylsilyl-	–	4·09	4·20
Acetamido- (MeCO·NH—)	3·61	4·03	
Phenoxy-	3·53	3·95	4·57
Acetyl-	4·13	3·83	3·70
Formyl-	–	3·84	3·77
Iodo-	2·86	3·87	4·00
Fluoro-	3·27	3·86	4·15
Chloro-	2·90	3·84	4·00
Bromo-	2·85	3·81	3·96
Cyano-	–	3·60	3·55
Aminosulphonyl- (—SO_2NH_2)	–	3·54	3·47
Nitro-	2·17	3·46	3·43
Phenyl-	3·46	–	–
Pentafluoro-		1·75	
2, 4-Dinitro-		1·43	
2, 4, 6-Trinitro-		0·65	

Naphthoic acids	1-	2-
	3·69	4·16

Heteroaromatic acids	2-	3-
Furan carboxylic	3·16	3·9[d]
Thiophene carboxylic	3·49	4·1[e]
Pyrrole carboxylic	4·45	5·00 (20°C)

[a]Brockman and Kilpatrick (1934), by conductimetry.
[b]King and Prue (1961), by potentiometry.
[c]Bolton, Fleming and Hall, (1972), by spectrophotometry.
[d]Catlin (1935).
[e]Voerman (1907).

a methyl-group in the *para* position even more than in the *meta* position. Again the nitro-, cyano-, and acetyl-groups have (− I) and (− M) properties, and hence the strength of benzoic acid is increased by these groups in the *para* position even more than in the *meta* position. In the more complicated cases where a group has both (− I) and (+ M) properties, the electron-withdrawing effects are less in the *para* than in the *meta* position, as is found for the hydroxy-, methoxy-, methylthio-, and halogen substituents. For the hydroxy- and methoxy-substituents, the (+ M) effect is so much more powerful than the (− I) effect that they actually have an acid-*weakening* effect in the *para* position (the hydroxybenzoic acids are to be found in Table 9.6).

The effects of several groups in the meta and para positions are almost numerically additive provided that strong (M) effects are absent.

The *ortho* position is the most complicated of all. In almost every set of isomers the *ortho* is the strongest acid, and even when a substituent has both (+ I) and (+ M) properties (e.g. the methyl-group), the resultant acid is stronger than benzoic acid. The *ortho* position must receive the strongest (I) effects in any set of isomers, because the distance is least. Again, like the *para* position, and unlike the *meta*, it is capable of propagating (M) effects. However, there is sometimes a third effect, which is acid-strengthening. This third effect is a steric pressure exerted by the substituent group upon the carboxyl-group, so that the latter is forced out of the plane of the ring. As a result of this distortion, the acid-weakening (+ M) effect of the benzene ring is decreased, i.e. the pK_a becomes referable to that of formic acid.

(c) *Aliphatic hydroxylic acids*

Many of the aliphatic hydroxylic acids ionize in a remote part of the pK scale where accurate results are elusive. For the aliphatic alcohols, we list in Table 9.3 a self-consistent set of results obtained conductimetrically by Ballinger and Long (1960), and preferred in the IUPAC compilation. Other results suggest that methanol and ethanol are not equally strong and, in any case, much weaker than pK 15·5 (see review by Bowden, 1966).

The aldehydes are monobasic acids because of their existence, in solution, as 1,1-glycols, e.g. $CH_2(OH)_2$ for formaldehyde. The latter, and chloral, are almost completely hydrated in solution, but aldehydes with (+ I) substituents (e.g. acetaldehyde) are only partly hydrated; hence the constant for acetaldehyde in Table 9.3 refers to the equilibrium:

$$\frac{[CH_3 \cdot CH(OH)O^-][H^+]}{[R \cdot CHO] + [R \cdot CH(OH_2)]}$$

Acetone has no detectable acidic properties, but enolizable ketones are highly acidic. The figures given in Table 9.3 are for the true enol. For acetylacetone and benzoylacetone, which enolize only slowly, total pK_a values of 8·94 and

Table 9.3   Aliphatic hydroxylic acids (alcohols, aldehydes and ketones)

In water. *Temperature*: 25°C, except where otherwise stated.

Acid	pK_a
Alcohols	
Methanol	15·5
Ethanol	15·5
Allyl alcohol	15·5
Ethylene glycol	15·1
2-Methoxyethanol	14·8
Propargyl alcohol $CH\vdots C\cdot CH_2OH$	13·6
2, 2, 2-Trifluoroethanol	12·37[a]
2, 2, 2-Trichloroethanol	12·24
Mannitol	13·5[b] (18°C)
Aldehydes and sugars	
Formaldehyde	13·27
Acetaldehyde	13·57
Trichloroacetaldehyde (chloral)	10·04
Pyridine-2-aldehyde	12·68
Pyridine-4-aldehyde	12·05 (30°C)
Benzaldehyde	14·90
Glutaconic dialdehyde $(OCH\cdot CH\vdots CH\cdot CH\vdots CHO^-)$	5·75
Ribose	12·22[c]
2-Deoxyribose	12·61
Glucose	12·46
Fructose	12·27
Mannose	12·08
Sucrose	12·7[d] (23°C)
Enolizable ketones	
Ethyl acetoacetate	10·68
Acetylacetone	9·0
Benzoylacetone	8·68
Triacetylmethane	5·01
Diacetylacetone	7·42
Ascorbic acid	4·04
1, 3-cyclohexanedione (dihydroresorcinol)	5·26

[a]See also Mukherjee and Grunwald (1958).
[b]Thamsen (1952).
[c]Except for sucrose, the values for these sugars were obtained by calorimetry (see Chapter 7).
[d]Stearn (1931).

8·70 respectively were first found, corresponding to the equilibrium:

$$\frac{[\text{enolate}^-][H^+]}{[\text{keto-form}] + [\text{enol-form}]}.$$

These figures were then adjusted for the percentage of enol present.

(d) Aromatic hydroxylic acids (phenols)

The increased acidic strength of phenol, as compared to the aliphatic alcohols, is usually attributed to the combined effect of the electron-attracting inductive effect $(-I)$ of the benzene ring, and the mesomeric effect $[(9·1) \leftrightarrow (9·2)]$ in the anion $(-M)$. As both effects have the same sign, the increase in acidic strength is large:

(9.1)   (9.2)

As in the benzoic acids, the *meta* position gives a clear picture of what inductive effects are operating. Thus the methyl-group is $(+I)$, and all other substituents shown in Table 9.4 are $(-I)$. The *para* position permits mesomeric, as well as inductive, effects. As before, where both effects have the same sign, the *para* position elicits a greater effect from the substituent than is possible in the *meta* position. For groups, like $-NO_2$, $-CN$ and $-SO_2CH_3$, which are $(-I, -M)$, striking lowering of pK_a values occurs, because the mesomeric potentialities of the phenol anion are so much greater than those of the benzoate anion, and the direction of the polarity is highly favourable. Where the (I) and (M) effects have opposite signs as in the halogens and in the methoxy-group, the relative strength of these effects can be gauged by the effect on acid strength. Thus F and, particularly, OMe, have powerful $(+M)$ effects, running counter to the $(-I)$ effect.

The *ortho* position is much less subject to steric hindrance of the kind noted in the substituted benzoic acids.

Phenol, the naphthols, and 1- and 2-anthrol all have very similar acid strengths.

Diphenols, phenolcarboxylic acids and phenol sulphonic acids are in Table 9.6. Aminophenols and the hydroxy-derivatives of heterocyclic bases are in Table 9.12.

(e) Other oxygen acids

Methane-sulphonic acid has a pK_a of $-1·20$, as determined by proton magnetic resonance (Covington and Lilley, 1967), cf. $-1·86$, as determined by Raman spectrometry (Clarke and Woodward, 1966); both values were obtained in water at 25°C. Thus this substance is comparable in strength to nitric acid

Table 9.4   Aromatic hydroxylic acids (phenols and naphthols)

In water. In order of increasing strength of meta-substituted isomer. *Temperature*: 25°C, except where otherwise stated (see Table 9.3 for aldehydes)

Acid	pK_a		
Phenol	9·99		
Substituted phenols	*o-*	*m-*	*p-*
t-Butyl-	10·62	10·12	10·23
Methyl- (cresols)	10·29	10·09	10·26
Phenyl-	10·01	9·64	9·55
Methoxy-	9·98	9·65	10·21
Acetamido- (—NH·COMe)	–	9·59	–
Aminocarbonyl- (—CONH$_2$)	8·89	9·30	8·56
Methylthio-	–	9·53	9·53
Acetyl-	10·06	9·19	8·05
Ethoxycarbonyl- (—CO$_2$Et)	9·92	9·10	8·34
Fluoro-	8·73	9·29	9·89
Chloro-	8·56	9·12	9·41
Bromo-	8·45	9·03	9·37
Iodo-	8·51	9·03	9·33
Formyl-	8·37	8·98	7·61
(Hydroxybenzaldehydes)			
Trifluoromethyl-	–	8·95	8·68
Methylsulphinyl- (—SOMe)	7·60	8·79	8·43
Methylsulphonyl-	–	8·40*	7·83
Cyano-	6·86	8·61	7·97
Nitro-	7·23	8·36	7·15
2, 4-Dinitro-		4·07	
2, 4, 6-Trinitro-		0·42	
Naphthols			
1-Naphthol		9·39	
2-Naphthol		9·63[†]	
5,6,7,8-Tetrahydro-		10·48	

*Bordwell and Cooper (1952), p. 6317, often misquoted as 9·33, p. 1058.
[†]This book, p. 52

(Table 9.13). Similarly p-toluenesulphonic acid was shown to have a pK_a of -1.34 (Dinius and Choppin, 1962) by pmr measurements in water at 25°C (cf. -1.06 for Raman measurement by Bonner and Torres, 1965). Some much lower values, e.g. -6.2 for methanesulphonic acid and -6.59 for p-toluene-sulphonic acid, which pmr has also furnished (Koeberg-Telder and Cerfontain, 1975), may refer to a very weakly basic oxygen atom, as in the equilibrium:

$$R \cdot SO_3H + H^+ \rightleftharpoons R \cdot SO_3HH^+.$$

Table 9.5 contains values for these sulphonic acids, and also for organic sulphinic and boric acids (see Table 9.6 for corresponding phosphorus and arsenic derivatives).

The remaining acids in this section contain nitrogen, but it is reasonable to suppose that the oxygen atom is the principal carrier of the negative charge. The acid amides are very weak acids (see Table 9.5), and urea has no detectable acidic properties. An N-substituent with a negative inductive effect strengthens

Table 9.5   Other oxygen acids (monobasic)

In water. *Temperature*: 25°C, except where otherwise stated.

Acid	pK_a	
Methanesulphonic acid[a]	-1.20	
p-Toluenesulphonic acid[a]	-1.34	
Benzenesulphinic acid	1.3	(20°C)
Benzeneboronic acid $[PhB(OH)_2]$	8.83	
2-Phenylethylboric acid	10.0	
Acetamide	15.1	
2,2-Dinitro-	1.30	(20°C)
Benzamide	$13-14$	
Succinimide	9.62	
Glutarimide[b]	11.43	
Acetohydroxamic acid (MeCO·NH·OH)	9.40	(20°C)
Benzohydroxamic acid	8.91	
Acetoxime (Me$_2$C·NOH)	12.42	
Nitromethane[a]	10.21	
Dinitromethane	3.57	
Trinitromethane	0.06	
Nitroethane	8.46	
Cyanic acid[c]	3.7	
*Tris*ethylsulphonylmethane[d]	< 1	

[a]See text.
[b]Schwarzenbach and Lutz (1940).
[c]Lister (1955).
[d]Samén (1947)

the acid properties of amides greatly. Some hydroxamic acids and oximes have also been included in Table 9.5.

For nitroparaffins, the equilibrium commonly measured is as (I), and constants for this are in Table 9.5. This is a slow ionization, involving the loss of a proton from a carbon atom (the negative charge finally rests on oxygen). Rapid back-titration reveals the equilibrium (II) for which nitroethane gives pK_a 4.4. The neutral isomer, shown in the lower line of (II), has only a short half-life.

$$\frac{[CH_3CH:NO_2^-][H^+]}{[CH_3CH_2NO_2]}$$

(I)

$$\frac{[CH_3CH:NO_2^-][H^+]}{[CH_3CH:NO_2H]}$$

(II)

9.2 The oxygen acids (dibasic)

When a symmetrical molecule with two identical ionizable groups is progressively neutralized, two well-separated constants are produced. In the first step of ionization, the equilibrium (III) can lose a proton from two positions, but it can add a proton to only one position. The second stage of neutralization (IV) can lose a proton from only one position, but can add it to two positions. Thus equilibrium (III) is statistically four times as favoured as equilibrium (IV).

(III) $HO_2C \cdot (CH_2)_4 \cdot CO_2H \rightleftharpoons HO_2C \cdot (CH_2)_4 \cdot CO_2^- + H^+$

(IV) $HO_2C \cdot (CH_2)_4 \cdot CO_2^- \rightleftharpoons {}^-O_2C \cdot (CH_2)_4 \cdot CO_2^- + H^+$

Because of the coulombic effect, the two constants of a symmetrical acid or base differ by *more* than this factor of four. Thus the negative charge in (III) must repel the similarly charged hydroxyl ions. Hence it is necessary to increase the concentration of hydroxyl ions (i.e. to increase the pH) very much more to achieve state (IV).

Table 9.6 gives examples of several dibasic acids, and it is evident that the difference between pK_a^1 and pK_a^2 varies greatly. Much depends on whether the two similar charges are near or distant, mutually hydrogen bonded or free. It also matters whether the coulombic effect is exerted through water (as in a long, thin molecule) or through the molecule itself (as in a short, thick molecule) because the dielectric constants are about 80 and 2, respectively. The extraordinary difference between pK_a^1 and pK_a^2 of salicylic acid is ascribed to hydrogen bonding of the first anion.

Activity correction for the *second* ionization constant is always large when two constants, for groups of the same charge-type, are measured potentiometrically. It is just as well to examine such results to make sure that the *full* correction (described in Chapter 3, p. 53) has been applied. Otherwise the second constant may really be that of an acid weaker than the published figures indicate.

Table 9.6   Polybasic oxygen acids

In water. *Temperature*: 25°C, except where otherwise stated.

Acid	More acidic pK_a	Less acidic pK_a
Aliphatic		
Oxalic	1·25	3·81
Malonic	2·85	5·70
Succinic[a]	4·21	5·64
Glutaric (18°C)	4·32	5·42
Adipic (18°C)	4·41	5·41
Oxaloacetic	2·55	4·37
Maleic	1·92	6·23
Fumaric	3·02	4·38
Tartaric (—)	3·03	4·37
Tartaric (meso)	3·17	4·91
Citric	3·13	4·76
		6·40
Aromatic		
Phthalic[b]	2·943	5·432
Isophthalic	3·70	4·60
Terephthalic	3·54	4·34
o-Dihydroxybenzene (catechol)	9·34	12·6
m-Dihydroxybenzene (resorcinol)	9·32	11·1
p-Dihydroxybenzene (hydroquinone)	9·85	11·4
o-Hydroxybenzoic (salicylic) (20°C)	2·98	13·6
m-Hydroxybenzoic	4·08	
p-Hydroxybenzoic	4·57	9·46
4-Sulphobenzoic ($HO_3S \cdot C_6H_4 \cdot CO_2H$)	–	3·65
3-Sulphophenol	–	9·07
4-Sulphophenol	–	9·11
Phosphorus and arsenic types		
Methyl dihydrogen phosphate [$MeO \cdot P(OH)_2O$]	1·54	6·31
Dimethyl hydrogen phosphate	1·25	–
Methanephosphonic acid [$MeP(OH)_2O$]	2·38	7·74
Benzenephosphonic acid	1·83	7·43
Benzenephosphinic acid (17°C)	2·1	–
Benzenearsonic acid[c]	3·47	8·49
Dimethylarsinic acid (cacodylic acid) [$Me_2As(OH)O$]	7·43	–

[a]From p.60.
[b]The mean (4·188) differs, because of activity corrections, from that (4·001) of Table 11.9.
[c]Pressman and Brown (1943), who give values for 18 other aromatic arsonic acids.

Table 9.7 Sulphur, nitrogen and carbon acids

In water. *Temperature*: 25°C, except where otherwise stated.

Acid	pK_a	
Sulphur acids		
Methanethiol (methyl mercaptan)	10·33	
2-Mercaptoethanol	9·72	
Phenylmethanethiol (benzyl mercaptan)	9·43	
Benzenethiol (thiophenol)	6·62	
Thioglycolic acid ($HS \cdot CH_2 \cdot CO_2H$)	3·68, 10·40	
Methyl thioglycolate (methyl 2-mercaptoacetate)	8·03	
Aminomethanedithioic acid (dithiocarbamic acid) ($H_2N \cdot CS_2H$)	2·95	
Methoxymethanedithioic acid (methylxanthic acid) ($MeO \cdot CS_2H$)	2·29	(0°C)
2-Mercaptoethanesulphonic acid	9·53	(20°C)
Nitrogen acids		
Aniline ($PhNH^-$)	27[a]	
2, 4, 6-Trinitroaniline[b]	12·2	
2, 4, 6-Trinitrodiphenylamine[b]	10·38	
Hexanitrodiphenylamine (dipicrylamine)[b]	2·63	
Cyanamide ($NC \cdot NH_2$)	10·3	
Pyrrole[c,d]	> 15	
Pyrazole[c,d]	∼ 14	
Imidazole[c,d]	14·4	
1, 2, 3-Triazole[c,d]	9·42	(20°C)
1, 2, 4-Triazole	10·26	
Tetrazole[e]	4·89	
Benzotriazole	8·6	
Purine[c,d]	8·93	(20°C)
8-Azapurine	4·84[e]	(20°C)
3-Nitrocarbazole[b]	14·1	
Carbon acids		
Diethyl malonate	12·9	
Malononitrile [$CH_2(CN)_2$]	11·20	
Cyanoform (tricyanomethane)	< 1	
Ethylacetoacetate[f]	–	
Nitromethane[g]	–	

[a]Conant and Wheland (1932), a value that is merely indicative.
[b]Stewart and O'Donnell (1962).
[c]For derivatives of these acids, see Albert (1963, 1971)
[d]For basic constants of these ampholytes, see Table 9.9.
[e]Albert (1966b). This acidic constant is little affected by the partial covalent hydration of the neutral species.
[f]See Table 9·3.
[g]See Table 9·5.

9.3 Sulphur acids, nitrogen acids and carbon acids

Apart from the many examples of acids where oxygen carries the negative charge, several types are known where this charge resides on sulphur, nitrogen, or carbon.

(a) *Mercaptans*

The high acidity of hydrogen sulphide (pK_a 7·05), compared to that of water, suggests that mercaptans are likely to be more acidic than the corresponding alcohols. This is borne out by the examples in Table 9.7. It is notable how much stronger an acid is thiophenol relative to phenol. For sulphonic and sulphinic acids, see Table 9.5.

(b) *Nitrogen acids*

The ready commercial availability of sodamide has made us familiar with the acidic nature of ammonia, but it is really very weak. Theoreticians have calculated a pK_a of about 33 for it. Nevertheless, the introduction of a strong electron-attracting group can lower this value enormously (see Table 9.7). In this connection, the low pK_a of nitramide $NO_2 \cdot NH_2$ (6.59) is relevant.

(c) *Carbon acids*

Carbanions are often invoked to explain organic chemical reactions, but few of them have more than a transitory existence. However, the substitution of two, or more, strongly electron-attracting substituents into methane gives stable acids, some of which are said to be as strong as mineral acids (see Table 9.7). For nitromethane, see Table 9.5. A number of hydrocarbons, roughly graded for acid strength by an indicator method, produced very high figures, such as indene (pK_a 21), fluorene (25), and triphenylmethane (33) and hence are exceedingly weak acids. Although these figures (Conant and Wheland, 1932) are only relative, it is notable that totally aromatic (six-membered ring) hydrocarbons were found to be even less acidic by the same test. For further reading on carbon acids, see Ebel (1969) and Jones (1973).

9.4 The nitrogen bases (monoacidic)

(a) *Aliphatic bases*

When ammonia (pK_a 9·3) is substituted by one alkyl-group, the pK_a rises to 10·6, a value which remains steady no matter how large the alkyl-group (see Table 9.8). When a second alkyl-group is added, further increase in basic strength is much smaller, and a third alkyl-group actually lowers the pK_a to a value intermediate between that of the secondary amine and ammonia. The increases in basic strength following the first two alkylations of ammonia are derived from the inductive effect ($+I$) of the alkyl-group. The decrease following the third alkylation (as in trimethylamine) has been attributed to

Table 9.8 Aliphatic bases (monoacidic)

In water. *Temperature*: 25°C, except where otherwise stated.

Base		pK_a
Amines		
Methylamine		10·66
dimethylamine		10·73
trimethylamine		9·80
Ethylamine		10·65
diethylamine		10·84
triethylamine		10·75
Propylamine		10·54
Isopropylamine ($Me_2CH \cdot NH_2$)		10·63
Butylamine		10·60
t-Butylamine ($Me_3C \cdot NH_2$)		10·68
Octylamine		10·65
Undecylamine		10·63
Dodecylamine		10·63
Hexadecylamine		10·61
Docosylamine ($C_{22}H_{45} \cdot NH_2$)		10·60
Substituted methylamines ($RCH_2 \cdot NH_2$)	R	
Phenyl- (benzylamine)	C_6H_5—	9·34
Aminocarbonyl- (glycinamide)	$NH_2 \cdot OC$—	7·95
Methoxycarbonyl- (glycine methyl ester)	$MeO \cdot OC$—	7·59
Vinyl- (allylamine)	$H_2C{:}CH$—	9·49
Cyano-	NC—	5·34
Substituted ethylamines ($RCH_2CH_2 \cdot NH_2$)	R	
Phenyl-	C_6H_5—	9·83
Methoxy-	MeO—	9·40
Hydroxy- (ethanolamine)	HO—	9·50
Ethoxycarbonyl-	$EtO \cdot OC$—	9·13[a]
Cyano-	NC—	7·80 (20°C)
Various		
Triethanolamine		7·76
Cyclohexylamine		10·64
Acetamidine[b]		12·40
Benzamidine (see Table 9.9)		
Trimethylamine-*N*-oxide		4·65 (20°C)
Heterocyclic bases (saturated)		
Piperidine		11·12
N-methyl-		10·38
Pyrrolidine		11·31
N-methyl-		10·46

Table 9.8 (*contd.*)

Base	pK_a	
Azetidine	11·29	
Aziridine	8·04	
Morpholine	8·50	
N-methyl-	7·38	
N-ethyl-	7·67	
Imines		
Cyclohexanimine (imine of cyclohexanone)	9·15	(?°C)
Diphenylketimine (iminobenzylbenzene)	7·35	(20°C)
4-Chlorobenzylidene-aniline (a Schiff base[c])	2·80	
Amides[d]		
Acetamide	− 1·4[e]	
Benzamide	− 2·16	
Urea	0·10	
O-Methylisourea $[H_2N \cdot C(:NH)OMe]$	9·72	(24°C)
Thiourea	− 1	
S-Methylisothiourea $[H_2N \cdot C(:NH)SMe]$	9·81	
Cyanamide $(NC \cdot NH_2)$	1·1	(29°C)
Acetylhydrazine	3·24	(24°C)
Oxygen bases		
Acetone	− 7·5[g]	(26°C)
Benzaldehyde	− 6·9[h]	
Acetophenone	− 7·6[g]	(26°C)
Benzophenone	− 8·4[h]	
Diethyl ether	_[f]	
Methanol, ethanol	_[f]	

[a]Edsall and Blanchard (1933).
[b]Guanidine and diguanide are in Table 9.10.
[c]Cordes and Jencks (1962). For anils derived from 4-aminoazobenzene, see Ricketts and Cho (1961).
[d]Acid amides are also oxygen bases: the oxygen atom takes up the hydrogen ion, whose charge is then shared between oxygen and nitrogen in a resonance hybrid.
[e]Deno and Wisotsky (1963), by Raman spectroscopy.
[f]See text p. 160.
[g]Levy, Cargioli and Racela, (1970).
[h]Greig and Johnson (1968).

'back-strain' overcrowding (Brown *et al.*, 1944); but the more widely held opinion is that progressive alkylation of an amine eventually reduces its strength by decreasing the number of hydrogen atoms (in the cation) capable of forming stabilizing hydrogen bonds with water (Bell, 1960).

    Tetra-alkylation, to give a quaternary salt, causes an immense increase in basic strength. In general, tetra-alkylammonium salts are completely ionized at all pH values; hence no pK_a values can be assigned to them. If the nitrogen atom, as is generally agreed, is incapable of forming five covalent bonds, no neutral species corresponding to a tetra-alkylammonium salt is conceivable. The low strength of ammonium salts, other than quaternary, is due to hydrogen-bonding to water molecules by lone pairs of electrons on the nitrogen atom in the neutral species. Tetra-alkylammonium salts do not possess any such lone pair.

The effects on pK_a of substituents on the α-carbon atom of methylamine, and the β-carbon atom of ethylamine, are shown in Table 9.8. The effects are all inductive and the substituents behave very much as with acetic acid (Table 9.1).

The value for allylamine raises the question of what would be the effect of moving the double-bond one place towards the nitrogen atom. Although simple aliphatic examples have not been measured, the phenomenon can be examined in the comparable tetrahydropyridines and dihydropyrroles (Adams and Mahan, 1942).

Amidines, thanks to extra resonance in the cations, are stronger bases than the simple aliphatic amines. At the other end of the scale are the amides which have only feebly basic properties. Guanidines are in Table 9.10, and hydrazines and hydroxylamines in Table 9.14.

Table 9.8 continues with a selection of heterocyclic bases, aliphatic in type (heteroaromatic bases are in Table 9.9).

The oxygen bases, with which Table 9.8 concludes, are discussed in Section 9.7, below.

(b) *Aromatic and heteroaromatic bases*

Aniline (pK_a 4·6) is a very much weaker base than either its saturated analogue (cyclohexylamine) or the aliphatic bases. This weakness is partly due to a resonance $[(9.3)\leftrightarrow(9.4)]$ in the neutral molecule which is not possible in the ion. However, a mere separation of opposite charges in a neutral molecule should not affect acidic or basic strength so markedly: moreover the dipole moment of aniline is only 1·6 D. Wepster (1952) presented evidence that about half of the base-weakening effect in aniline comes from the inductive effect of the benzene ring which has the same sign $(-I)$ as the above resonance effect $(-M)$. In benzoic acid, on the other hand, the two effects have opposite signs.

(9.3) (9.4)

The pK_a values of many derivatives of aniline are given in Table 9.9. It will be noted that *N*-ethyl substituents increase basic strength by more than could

Table 9.9 Aromatic and heteroaromatic bases (monoacidic)
In water. *Temperature*: At 25°C, except where otherwise stated

Base	pK_a		
Aniline	4·87		
N-Methyl-	4·85		
NN-Dimethyl-	5·07		
N-Ethyl-	5·12		
NN-Diethyl-	6·57		
N-Isopropyl-	5·77		
N-t-Butyl-	7·00		
N-Phenyl-(diphenylamine)	0·79		
N-Acetyl- (acetanilide)	∼0·5		

C-*Substituted anilines*	*o-*	*m-*	*p-*
Methyl- (toluidines)	4·45	4·71	5·08
Methoxy- (anisidines)	4·53	4·20	5·36
Phenyl- (18°C)	3·83	4·25	4·35
Methylthio- (MeS-)	3·45	4·00	4·35
Fluoro-	3·20	3·59	4·65
Chloro-	2·66	3·52	3·98
Bromo-	2·53	3·53	3·89
Iodo-	2·54	3·58	3·81
Methoxycarbonyl- (MeO·OC—)	2·32	3·55	2·47
Trifluoromethyl-	–	3·49	2·45
Trifluoromethylthio-	–	3·30	2·78
Trifluoromethoxy-	2·44	3·25	3·82
Cyano	0·77	2·75	1·74
Methylsulphonyl- (MeO$_2$S—)	–	2·58	1·36
Nitro-	−0·25	2·46	1·02
Trifluoromethylsulphonyl-	–	1·79	−0·01

Various			
2, 6-Dimethylaniline	3·89		
NN-Diethyl-o-toluidine	7·24		
1-Naphthylamine	3·92		
2-Naphthylamine	4·16		
Benzamidine	11·6 (20°C)		
Methyl benzimidate [PhC(:NH)OMe]	5·8		
Pyridine	5·23		
2-Methyl	6·00		
3-Methyl-	5·70		
4-Methyl-	5·99		
2-Methoxy-	3·28 (20°C)		
3-Methoxy-	4·78		
4-Methoxy-	6·58		
3-Acetyl-	3·26		
2-Fluoro-	−0·44		
2-Chloro-	0·49		

Table 9.9 (*contd.*)

Base	pK_a	
3-Chloro-	2·81	
4-Chloro-	3·83	
3-Cyano-	1·35	
4-Nitro-	1·61	
2-Amino-	6·82	(20°C)
3-Amino-	6·04	
4-Amino-	9·11	
4-Methylamino-	9·65	
N-Oxide	0·79	(24°C)
Other heteroaromatic bases		
Quinoline	4·90	(20°C)
2-amino	7·34	(20°C)
4-amino-	9·17	(20°C)
Isoquinoline	5·40	(20°C)
1-amino-	7·62	(20°C)
3-amino-	5·05	(20°C)
Acridine	5·58	(20°C)
9-amino-	9·99	(20°C)
3, 6-diamino-	9·65	(20°C)
Pyridizine	2·24	(20°C)
Pyrimidine	1·23	(20°C) (also − 6·9)
4-amino-	5·71	(20°C)
2, 4, 6-triamino-	6·84	(20°C)
Pyrazine	0·65	(20°C) (also − 6·3)
Phthalazine	3·47	(20°C)
Cinnoline	2·37	(20°C)
Quinazoline	1·94[a]	(29°C)
	3·43[b]	(29°C)
Quinoxaline	0·56	(20°C)
Phenazine	1·20	(20°C)
Pyrrole[c]	− 3.8	
Pyrazole[c]	2·49	
Imidazole[c]	6·99	
1, 2, 3-Triazole[c]	1·17	(20°C)
1, 2, 4-Triazole[c]	2·27	(20°C)
Tetrazole[c]	−[d]	
Benzimidazole	5·53	
Benzotriazole[c]	1·6	(20°C)
Purine[c]	2·30	(20°C)
Thiazole	2·52	
Oxazole	0·8	(33°C)
Isoxazole	− 2·0	

[a] pK_a for metastable equilibrium between anhydrous cation and anhydrous molecule, determined in rapid-reaction apparatus.
[b] pK_a for stable equilibrium between hydrated cation and anhydrous molecule.
[c] See Table 9·7 for ionization of the acid function.
[d] No basic pK_a above 0.

be expected from their (+ I) effect. They are, by their bulk, twisting the amino-group out of the plane of the benzene ring, a process which interferes with the base-weakening resonance $[(9.3)\leftrightarrow(9.4)]$. That this interpretation is correct is confirmed by the greater basic strength of *N-tert*-butylaniline ($pK_a = 7\cdot0$). Alkyl substituents, if placed *ortho* to the amino-group, lower basic strength by hindering the approach of the hydrated proton, cf. 2-*tert*-butylaniline, $pK_a = 3\cdot8$, and 2, 4, 6-tri*tert*-butylaniline, $pK_a < 2$ (Bartlett *et al.*, 1954).

In general, *meta*-substituted anilines show the usual inductive effects, already discussed in connection with aromatic carboxylic acids (p. 140). These effects usually influence the magnitude of the pK_a of aniline even more than they influence that of phenol (Table 9.4). In the *para* position, these inductive effects are reinforced by any mesomeric effect of which the substituent is capable, as explained on p. 140. The halogens and the methoxy-group show the usual resultant of (− I) and (+ M) effects.

Table 9.9 also records the pK_a values of benzamidine, diphenylamine, acetanilide, the naphthylamines, and a small selection of heteroaromatic bases. It will be noted that a second ring-nitrogen atom depresses basic strength (by a (− I) effect) except in a few cases where a base-strengthening resonance counteracts this effect (e.g. pyrazole, imidazole).

It will be seen from Table 9.9 that, in general, the effect of a substituent in pyridine is similar to its effect in aniline. However, there is almost no *ortho*-effect in pyridine because the cation-forming atom does not project beyond the ring. The 2- and 4-aminopyridines, also similarly substituted quinolines and acridines, are unusually strong bases because of a resonance in the cation $[(9.5)\leftrightarrow(9.6)]$ that is not possible in the neutral species (Albert, Goldacre and Phillips, 1948). For listings of values for purines and pyrimidines, see Albert (1973); and for other heterocycles see Albert (1963, 1971).

9.5 The nitrogen bases (diacidic)

The relationship between the pK_a values of two basic groups in the same molecule is governed by the influences that control the relationship between two acidic groups in one molecule (see p. 147). The need for another activity correction for the second constant (outlined on p. 53) applies here, too.

Several examples of diacidic bases are given in Table 9.10. The aliphatic bases in this table clearly show how an ionized amino-group has a much larger effect (than a non-ionized amino-group) on the ionization of the other amino-group. These effects fall off as the two amino-groups are separated by more carbon atoms, but the effect of the *ionized* amino-group persists further. Hence,

Table 9.10 Diacidic bases

In water. *Temperature*: At 25°C, except where otherwise stated.

Base	More basic pK_a	Less basic pK_a
Aliphatic		
1, 2-Diaminoethane	9·92	6·86
1, 3-Diaminopropane	10.55	8.88
1, 4-Diaminobutane	10.80	9.63
1, 8-Diaminooctane (20°C)	11·00	10·1
1, 2-Diaminocyclohexane (20°C)		
cis-	9·93	6·13
trans-	9·94	6·47
1, 3-Diaminopropan-2-ol (20°C)	9·69	7·93
Piperazine	9·73	5·33
Guanidine	13·6	− 11*
N-acetyl- (23°C)	8·20	
N-methyl-	13·4	
Diguanide	11·49	2·95
Aromatic		
Benzidine† (20°C)	4·65	3·43
1, 2-Diaminobenzene		
(*o*-phenylenediamine (20°C)	4·57	0·80
1, 3-Diaminobenzene (20°C)	5·11	2·50
1, 4-Diaminobenzene (20°C)	6·31	2·97

*Williams and Hardy (1953).
† This book, Table 4.8

in a homologous series, the bigger differences are found in the second ionization step (i.e. of the weaker basic group).

The aromatic diamines present a slightly more complex picture because a base-strengthening (+ M) effect is transmitted from a non-ionized amino-group in the *ortho* and *para* positions. In the *ortho* position, this is counteracted by the steric effect (p. 142) which is base-weakening.

9.6 Carbinolamine bases

This heading relates to several examples where the cation and the molecule have quite different chemical structures, the molecule being derived from the cation by a nucleophilic attack of a hydroxyl anion during titration with alkali. This produces an alcohol (the carbinolamine). Heteroaromatic quaternary amines provide typical examples. It might be expected that these would be ionized completely at all pH values, just as tetramethylammonium chloride is. However, the presence of a double-bond on the tetracovalent nitrogen atom confers a special character on such structures, so that we speak of the 1-methyl-pyrid*inium* (not pyrid*onium*) ion. This substance, the simplest member of the series, is not susceptible to nucleophilic attack by the hydroxyl ion, but becomes so when substituted by electron-attracting groups. Thus 3, 5-dicyano-1-methylpyridinium chloride (*9.7*) is in rapidly reversible equilibrium with

Table 9.11   Carbinolamines

At equilibrium in water.

Temperature: 25°C, except where otherwise stated

Parent substance	$pK_{equil.}$	°C
Quaternary nitrogen heteroaromatics		
1-Methyl-3, 5-dicyanopyridinium	3·5[a]	
1-Methylquinolinium	~ 16·5[b]	
3-nitro-	6·74[a]	
4-nitro-	5·31[b]	
2-Methylisoquinolinium	~ 15·3[b]	
10-Methylacridinium[c]	9·85[d] ⎫	
3-amino-	12·02[d] ⎬ 20	
9-amino-	11·18[d] ⎭	
3, 6-diamino (euflavine, acriflavine)	> 12	
5-Methylphenanthridinium	10·4[a]	
1-Methylquinoxalinium	8·62[b]	
2-Methylphthalazinium	11·04[b]	
1-Methyl-1, 5-naphthyridinium	12·67[b]	
3, 5-Dimethylthiazolium	10·8[a]	
3-Methylbenzothiazolium	8·30[a]	
Tertiary oxygen or sulphur heteroaromatics		
2, 4, 6-Trimethylpyrylium	6·7[a]	
Benzopyrylium	− 1·96[a]	
Benzothiopyrylium	3·15[a]	
Triphenylmethane dyes		
Triphenylcarbinol		
4, 4′-diamino- (Doebner's violet)	5·38[d]	
4, 4′, 4″-triamino- (Parafuchsin)	7·57[d]	
4-Dimethylamino-	4·75[d]	
4, 4′-bisdimethylamino- (Brilliant green)	7·90[d]	
4, 4′-bisdiethylamino- (Malachite green)	6·90[d]	
4, 4′, 4″-trisdimethylamino- (Crystal violet)	9·36[d]	

[a]Bunting (1979).
[b]Bunting and Meathrel (1972).
[c]The hydroxyl-ion attacks in the 9-position (Albert, 1966a).
[d]Goldacre and Phillips (1949).

3, 5-dicyano-2-hydroxy-1-methyl-1, 2-dihydropyridine (*9.8*), and a potentiometric titration yields a pK_a of 3·5 (Table 9.11). This figure, better denoted pK_{equil}, is the resultant of two processes, (a) the ionization of the carbinolamine and (b) the tautomerism of this ion to 3, 5-dicyano-1-methylpyridinium hydroxide.

If the pyridine ring is annelated to one or more six-membered rings, or if more doubly-bound nitrogen atoms are present, the transition to carbinolamines becomes much easier, as the lower values in Table 9.11 show. The subject has been well reviewed by Bunting (1979).

Ring-opening does not follow carbinol formation in completely aromatic rings, but is common in partially hydrogenated structures such as the 3, 4-dihydroquinolines (Beke, 1963). Five-membered rings containing both a nitrogen and an oxygen (or sulphur) atom behave similarly, e.g. 3-methylthiazole readily takes part in ring-opening after attack by a hydroxyl ion in the 2-position. Such valence-isomerism is readily reversed on back-titration.

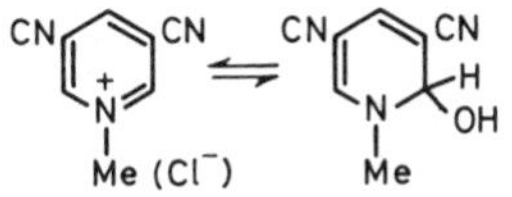

(*9.7*)   (*9.8*)

Those six-membered rings in which oxygen or sulphur is the sole heteroatom, also undergo attack by a hydroxyl ion during titration. Thus pyrylium cation (*9.9*) enters into easily reversible equilibrium with 2-hydroxypyran (*9.10*). The latter then spontaneously undergoes ring-opening to penta-4-ene-dialdehyde, no ionic catalysis being required (thermal reaction). The product reverts to pyrylium chloride on back-titration. Some examples of pK_{equil} derived from the pyrylium ring can be found in Table 9.11. Once again, Bunting's review provides a good discussion and further examples.

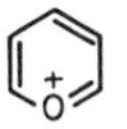

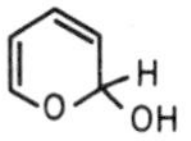

(*9.9*)   (*9.10*)

A different, but related, type of carbinolamine formation is characteristic of the diamino- and triamino-triphenylmethane dyes, e.g. parafuchsin (pararosaniline) (*9.11*). Here, transition to the carbinolamine form (*9.12*) is often

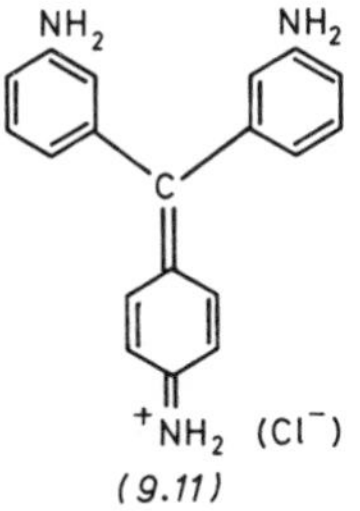

(*9.11*) 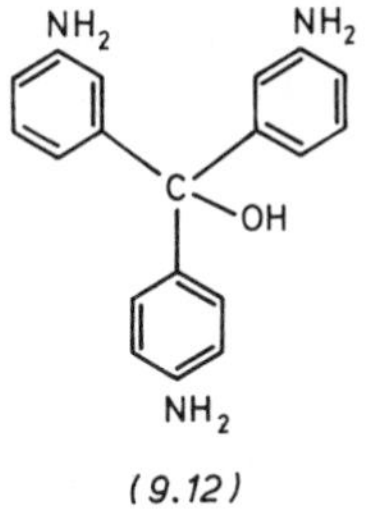   (*9.12*)

slower than with the heterocycles, but yields a similar pK_{equil}. Several examples will be found in Table 9.11. The correct treatment of this ionization was first worked out by Goldacre and Phillips (1949), who give several further examples.

Typical half-times $(T_{0.5})$ for the equilibrium, in hours, are: crystal violet (7·5), brilliant green (2·9), and parafuchsin (0.85).

9.7 Oxygen bases and carbon bases

Basic properties can originate in atoms other than nitrogen. Some oxygen bases are listed at the end of Table 9.8. There is general agreement about the pK_a for acetone. The value ($-7·5$), chosen for Table 9·8 is taken from PMR measurements by Levy, Cargioli and Racela, (1970) and supersedes similar values (both of them $-7·2$) obtained by ultraviolet spectrophotometry (Edward, Chang, Yates and Stewart, 1960), and by Raman spectroscopy (Deno and Wisotsky, 1963).

Little agreement exists about the basic strength of diethyl ether for which the value $-6·2$ has been found by PMR (Edward, Leane and Wang, 1962) but $-3·6$ from partition studies (Arnett and Wu, 1960), also a similar value ($-3·53$ from vapour-pressure measurements at varied H_0 values (Jaques and Leisten, 1964), and the lower $-2·42$ from more recent PMR studies (Bonvicini *et al.*, 1973).

The primary alcohols have not fared much better, particularly as they become esterified in strongly acidic solutions, whereas ether is quite stable. For methanol, the value $-2·2$ was obtained by Raman spectroscopy (Deno and Wisotsky, 1963) but was rejected in favour of $-4·8$ by Weston, Ehrenson and Heinzinger, (1967) who detected (and avoided) a likely source of error in the earlier work. However, more recently, the value $-1·98$ has been preferred by Bonvicini *et al.*, who used PMR. Ethanol has been examined, at 0°C to minimize esterification, by Lee and Cameron (1971) who reported the value of $-1·94$ which they prefer to $-4·8$ obtained (at 23°C) by Edward, Leane and Wang, (1962). Thus it is clear that diethyl ether and the primary alcohols are weak bases, but we are still unsure as to how weak they really are.

At pK_a values so low as these, it is possible that protonation is taking place on carbon in some cases, just as happens in pyrrole (Chiang and Whipple, 1963). Choice of the correct acidity function, not necessarily H_0, is important, and has been discussed by Greig and Johnson (1968).

Some oxygen heterocycles are more basic than the aliphatic examples in Table 9.8, because extra resonance is possible in the (heterocyclic) cations. Thus 4-pyrone and anthocyanidin have pK_a values of $+0·3$ and $4·0$, respectively.

A rare example of a stable carbonium ion is provided by 2, 4, 2′, 4′, 2″, 4″-hexamethoxytriphenylcarbinol which has a pK_a of 3·2 at 15°C (Kolthoff, 1927).

9.8 Amphoteric substances

The discussion on amphoteric substances in Chapter 8 forms a suitable introduction to Table 9.12 which contains a selection of typical examples.

Table 9.12   Amphoteric substances

In water. *Temperature*: 25°C except where otherwise stated

Substance	pK$_a$		
	Proton gained	Proton lost	°C
Aliphatic			
2-Aminoacetic acid (glycine)*	2·3503	9·7796	
2-Aminopropionic acid (alanine)*	2·34	9·87	
3-Aminopropionic acid	3·55	10·24	
5-Aminovaleric acid	4·27	10·77	
2-Mercaptoethylamine	8·27	10·53	
Aminomethanesulphonic acid	–	5·75	
Aminoethane-2-sulphonic acid (taurine)	–	9·06	
Aromatic			
2-Aminophenol	4·78	9·97	20
3-Aminophenol	4·37	9·82	20
4-Aminophenol	5·48	10·30	
2-Aminobenzoic acid (anthranilic acid)	2·17	4·85	
3-Aminobenzoic acid[†]	3·07	4·79	
4-Aminobenzoic acid	2·50	4·87	
Aniline-2-sulphonic acid	–	2·46	
Aniline-3-sulphonic acid	–	3·74	
Aniline-4-sulphonic acid (sulphanilic acid)	–	3·23	
Heteroaromatic			
Pyridine-2-carboxylic acid (picolinic acid)	0·99	5·39	
Pyridine-3-carboxylic acid (nicotinic acid)	2·00	4·82	
Pyridine-4-carboxylic acid	1·77	4·84	
2-Pyridone (2-hydroxypyridine) ،	0·75	11·65	20
3-Hydroxypyridine	4·79	8·75	20
4-Pyridone	3·20	11·12	20
2-Quinolone	− 0·31	11·76	20
3-Hydroxyquinoline	4·28	8·08	20
4-Quinolone	2·23	11·28	20
6-Hydroxyquinoline	5·15	8·90	20
8-Hydroxyquinoline (oxine)	4·91	9·81	
2-Mercaptoquinoline	− 1·44	10·25	20
3-Mercaptoquinoline	2·29	6·17	20
4-Mercaptoquinoline	0·77	8·87	20

*For other aminoacids, see Table 9.15.
[†]See Table 4.7.

Most of the aminoacids derived from protein have low pK_a values about 2·0 and upper values between 8 and 10 (some have two upper values and are basic whereas others have two lower values and are acidic). See Table 9.15 for pK_a values of these aminoacids.

The pK_a values of thirteen naphthylamine-sulphonic acids (proton lost constants only) are available to supplement the naphthylamine values of Table 9.9 (Bryson, 1951).

B   INORGANIC SECTION

9.9 Inorganic acids

A number of inorganic acids are listed in Table 9.13. All the values relate to water, but some of the extreme figures (e.g. NH_3 and HI) have been obtained by theoretical estimation only and even the order of magnitude may have to be revised. In any one column of the periodic table, acid strength increases

Table 9.13   Inorganic acids

In water. *Temperature*: 25°C, except where otherwise stated; a dash (–) implies that authors published no temperature

Acid	pK_a	°C
Amidophosphoric acid	2·74, 8·10	
Aminodisulphonic acid	8·50	
Ammonia (NH_2^-)	~ 33[a]	–
Aluminic acid	11·2	–
Arsenic acid	2·22, 6·98	
Arsenious acid	9·18	
Boric acid	9·234	
(Carbonic acid)	6·35, 10·33	
Chlorous acid	1·94	
Chromic acid	0·74, 6·49	–
Cyanic acid	3·46	–
Diamidophosphoric acid	1·23, 4·94	
Diamidothiophosphoric acid	2·0, 4·3	20
Ferricyanic acid	< 1	–
Ferrocyanic acid	2·57, 4·35	
Fluorophosphoric acid	5·12	
Germanic acid	8·73, 11·7	
Hydrazinosulphuric acid	3·85	–
Hydrazoic acid	4·72	
Hydriodic acid	– 9	–
Hydrobromic acid	– 8	–
Hydrochloric acid	– 6·1	

Table 9.13 (*contd.*)

Acid	pK_a	°C
Hydrocyanic acid	9·22[b]	
Hydrofluoric acid	3·18	
Hydrogen peroxide	11·65	
Hydrogen selenide	3·8	
Hydrogen sulphide	7·05, 14·0	20
Hypobromous acid	8·49	
Hypochlorous acid	7·54	
Hyponitrous acid	7·15, 11·54	
Hypophosphorous acid	1·23	
Iodic acid	0·80	
Molybdic acid	3·57, 4·84	
Nitramide	6·48	
Nitric acid	− 1·44	
Nitrous acid	3·20	20
Oxyhyponitrous acid	2·51, 9·70	1
Perchloric acid[c]	–	–
Periodic acid	1·64, 8·36	
Permanganic acid	− 2·25	
Phosphoric acid	2·148, 7·198, 12·375	
Deuterophosphoric acid	2·35, 7·78	
Phosphorous acid	1·43, 6·67	
Plumbic acid (from Pb^{2+})	11·5	
Pyrophosphoric acid	1·52, 2·36, 6·60, 9·25	
Selenic acid	1·66	
Selenious acid	2·62, 8·32	
Silicic acid	9·77	
Sulphamic acid	0·99	
Sulphuric acid	− 3, + 1·96	
Deuterosulphuric acid	2·33	
Sulphurous acid	1·89, 7·21	
Telluric acid	7·70, 10·95	
Tetracyanonickelic acid	4·7, 6·6	
Thiocyanic acid	− 1·85	
Thiophosphoric acid (H_3PO_3S)	1·8, 5·4, 10·1	
Thiosulphuric acid ($H_2S_2O_3$)	1·74	
Trithiocarbonic acid	2·7, 8·2	20
Vanadic acid	3·78	

[a]Pleskov and Monoszon (1935); this is merely indicative. It was found at − 50°C, in absence of water.
[b]Ang (1959), by indicator method.
[c]Completely ionized, even in 10M solution.

with rising atomic number, at least for the simple hydrides (cf. H_2O, H_2S, H_2Se, and H_2Te; again, cf. HF, HCl, HBr, HI). However, acid strength is not *closely* connected with electronegativity.

The strengths of the mono-anions of oxy-acids increases with the number of oxygen atoms *not* linked to hydrogen. When there is only *one* such atom, the pK_a lies between 7 and 11. When there are *two* such oxygens, the pK_a is between 1 and 3·5; similarly *three* such oxygen atoms give values between pK_a − 3 and − 1. Perchloric acid, which seems to be the strongest of all inorganic acids, has four such atoms. The di-anions fall on a parallel scale. The values for di-anions are higher, because the coulombic effect of one negative charge hinders the approach of hydroxyl ions, and hence makes the second ionization more difficult. Phosphorous and hypophosphorous acid fit the 'oxygen rule' because they both have pentavalent phosphorus.

Some of the ionization constants in Table 9.13 are only resultants of a true ionization constant and a dehydration constant (e.g. $HCO_3^- \rightleftharpoons H_2CO_3 \rightleftharpoons CO_2$, where the true pK_a is 3·8; i.e. it is about 400 times stronger than a simple titration reveals). Periodic acid and telluric acid are also much weaker than would be expected, because of hydration phenomena.

The temperature coefficients of inorganic acids differ enormously, but are usually large.

9.10 Inorganic bases

The pK_a values of some inorganic bases are listed in Table 9.14. They are expressed as pK_a values by following the concepts of Werner (1907) and Pfeiffer (1907) that protons are derived from water molecules bound to metal ions. Thus the ionization of potassium hydroxide is governed by this equilibrium:

$$K_a = \frac{[H^+][KOH]}{[K^+, H_2O]} \tag{9.1}$$

where K_a is termed the 'hydrolysis constant'.

It is true that we are usually more aware of the hydroxyl- than of the hydrogen-ion concentration in potassium hydroxide. However, equation (9.1) is formally similar to that used for organic bases, and can be used for inorganic bases in the generalized form:

$$K_a = \frac{[H^+][MOH_{aq}^{(n-1)+}]}{[M_{aq}^{n+}]}. \tag{9.2}$$

It is not easy to select, from a voluminous literature, many values that truly represent this simple equilibrium. Heavy metals often form polynuclear complexes, which involve new ionic species and complicate the equilibria. Thus the pK_a of cupric ion has been reported as 6·5 to 7·9, but meticulous study has shown that polynuclear complexes, especially $HO \cdot Cu^+ - Cu^+ \cdot OH$, are the main oxygen-containing species present (Pedersen, 1943; Perrin, 1960). At

Table 9.14 Inorganic bases

In water. *Temperature*: 25°C, except where otherwise stated; a dash (−) implies that authors published no temperature

Base	pK_a	°C
Aluminium (Al^{3+})	5	−
Ammonia	9·246	
deuterio-ammonia	9·757	20
bromamine	6·39	−
Barium	13·36	
Bismuth (Bi^{3+})	1·58	
Cadmium	9–10	−
Calcium	12·9	
Chromium (Cr^{3+})	4	−
Cobalt (Co^{2+})	9·85	
Copper[a]	−	−
Hydrazine	− 0·88, + 8·11	20
methyl-	7·87	30
tetramethyl-	6·30	30
phenyl-[b]	5·27	
Hydroxylamine[c]	5·96	
O-methyl[d]	4·60	
N, N-dimethyl[d]	5·20	
trimethyl[d]	3·65	
Iron (Fe^{2+})	6·74	
Iron (Fe^{3+})	2·46	
Lead (Pb^{2+})	8	−
Lithium	13·82	
Magnesium	11·42	
Manganese (Mn^{2+})	10·59	
Mercury (Hg^+)	5·0	
Mercury (Hg^{2+})	3·49	
Nickel	9·86	
Palladium (Pd^{2+})	1	
Potassium	16	
Scandium	4·55, 8·8	
Silver	> 11	
Sodium	14·77	
Strontium	13·2	
Thallium (Tl^+)	13·2	
Thallium (Tl^{3+})	1·16	
Tin (Sn^{2+})	1·70	
Uranyl ion (UO_2^{2+})	5·82	
Zinc	8·96	

[a]The polynuclear complexes, which Cu^{2+} readily forms, have so far prevented determination of a pK_a.
[b]Stroh and Westphal (1963); many substituted phenylhydrazines in same paper.
[c]Robinson and Bower (1961).
[d]Bissot *et al.* (1957).

the other extreme, mercuric ion behaves in a perfectly orthodox way, giving $HOHg^+$ and $(HO)_2Hg$ with no signs of any polynuclear complex (Hietanen and Sillén, 1952). Ferric ion occupies an intermediate position. It is emphasized that such studies must be made with a non-complexing anion, preferably perchlorate.

C   BIOLOGICALLY-ACTIVE SUBSTANCES

This chapter ends with a table of substances whose ionization constants are likely to interest many people who work in biological, medicinal, or pharmaceutical fields. Table 9.15 contains values for aminoacids (Perrin, 1965,

Table 9.15   Substances of interest in biology, medicine and pharmacy

Notation of the type: (T 9.2) refers to tables in this chapter

Substance	pK_a (approx.)
Acetic acid (T 9.1)	
Acetylcholine (ionized at all pH values)	
Acriflavine (T 9.11)	
Acyclovir	2·3, 9·3
Adenine	4·3, 9·8
Adenosine	3·6, 12.4
Adenylic acid (AMP)	3·7, 6·1, 13·1
Adrenaline (epinephrine)	8·7, 10·2, 12·0
Adriamycin	8·2, 10·2
Allopurinol	9·4
Alprenolol	9·5
Amantidine	10·1
Amethocaine (tetracaine)	8·5
Amidopyrine	5·1
Amiloride	8·7
Aminacrine (T 9.9)	
Aminobenzoic acid (T 9.12)	
4-Aminobutyric acid (GABA)	4·0, 10·6
Aminocaproic acid (6-Aminohexanoic acid)	4·4, 10·8
Aminocephalosporanic acid	2·0, 4·4
p-Aminosalicylic acid	1·8, 3·6
Amitriptyline	9·4
Amoxycillin	2·4, 7·4, 9·6
Amphetamine	9·9
Amphoteracin	5·5, 10·0
Ampicillin	2·5, 7·3
Amylobarbital	7·9
Antazoline	2·5, 10·1

Table 9.15 (*contd.*)

Substance	pK_a (approx.)
Apomorphine	7·2, 8·9
Arecoline	7·4
Arginine	1·8, 9·0, 12·5
Ascorbic acid (T 9.3)	
Aspirin	3·5
Asparagine	2·1, 8·7
Aspartic acid	2·0, 3·9, 10·0
Atebrin (*see* Mepacrine)	
Atenolol	9·6
Atropine	10·2
Baclofen	3·9, 9·6
Bemegride	11·6
Bendrofluazide	8·5
Benzocaine	2·5
Benzoic acid (T 9.2)	
Benztropine	10·0
Benzylpenicillin (Penicillin G)	2·8
Boric acid (T 9.13)	
Brilliant Green (T. 9.11)	
Bromocriptine	4·9
Butobarbital	9·0
Cacodylic acid (T 9.6)	
Caffeine	< 1
Carbenoxolone	6·7
Carbonic acid (T 9.13)	
Cephalexin	5·2, 7·3
Cephaloridine	1·7, 3·4
Cephalothin	2·2
Cephradine	2·5, 7·3
Chloral hydrate (T 9.3)	
Chlorambucil	5·8, 8·0
Chloramphenicol	5·5
Chlordiazepoxide	4·6
Chlorocresol (4-chloro-3-methylphenol)	9·2
Chloropyrilene	8·4
Chloroquine	8·4, 10·8
Chlorothiazide	6·7, 9·5
Chlorpheniramine	4·0, 9·2
Chlorphentermine	9·6
Chlorpromazine	9·3
Chlorpropamide	5·0
Chlortetracycline	3·3, 7·4, 9·3
Cimetidine	6·8 (ring)

Table 9.15 (*contd.*)

Substance	pK_a (approx.)
Cinchocaine (dibucaine)	7·5
Citric acid (T 9.6)	
Clindamycin	7·7
Clofibrate	3·0
Clonidine	8·2
Cloxacillin	2·7
Cocaine	8·4
Codeine	8·2
Colchicine	1·9
Cresols (T 9.4)	
Cromolyn sodium	2·5
Crystal violet (T 9.11)	
Cyclobarbital	7·6
Cyclopentolate	7·9
Cycloserine	4·5, 7·4
Cysteine	1·5, 8·7, 10·2
Cytarabine	4·3
Cytidine	4·1, 12·2
Cytosine	4·6, 12·1
Dacarbazine	4·4
Dantrolene	7·5
Dapsone	1·3, 2·5
Daunorubicin	8·2
Debrisoquine	11·9
Demeclocycline	3·3, 7·2, 9·2
Desipramine	10·2
Dexamphetamine	9·9
Dextromethorphan	8·3
Dextrorphan	4·8
Diamorphine (heroin)	7·6
Diatrizoic acid	3·4
Diazepam	3·3
Dicloxacillin	2·7
Dicoumarol	4·4, 8·0
Diethylcarbamazine	7·7
Dihydrocodeine	8·8
Dihydroergotamine	6·9
Diodone	2·8
Diphenhydramine	9·0
Dipipanone	8·5
Dopamine	8·8, 10·6
Doxorubicin (adriamycin)	8·2, 10·2
Doxycycline	3·5, 7·7, 9·5

Table 9.15 (*contd.*)

Substance	pK_a (approx.)
Emetine	7·6, 8·4
Epinephrine (*see* adrenaline)	
Ephedrine	9·6
Ergometrine	6·8
Ergotamine	6·4
Erythromycin	8·9
Ethacrynic acid	3·5
Ethambutol	6·3, 9·5
Ethopropazine	9·6
Ethosuximide	9·5
Ethylenediaminetetracetic acid (EDTA)	2·0, 2·7, 6·2, 10·3
Fenfluramine	9·1
Flufenamic acid	3·9
Fluopromazine	9·2
Fluorouracil	8·0, 13·0
Fluphenazine	3·9, 8·1
Formaldehyde (T 9·3)	
Frusemide	3·9
Fusidic acid	5·4
Gentamicin	8·2
Glibenclamide	5·3
Glucose (T 9·3)	
Glutamic acid	2·2, 4·2, 9·6
Glutamine	2·2, 9·0
Glutethimide	4·5, 9·2
Guanethidine	8·3, 11·4
Guanosine	3·0, 9·3, 12·6
Haloperidol	8·3
Heroin	7·6
Hexobarbital	8·2
Histamine	5·9, 9·7
Histidine	1·8, 6·0, 9·3
Homatropine	9·7
Hydrallazine	7·0
Hydrochlorothiazide	7·9, 9·2
Hydroflumethiazide	8·5, 10·0
Hydrogen peroxide (T 9·13)	
8-Hydroxyquinoline (oxine) (T 9·12)	
5-Hydroxytryptamine (serotonin)	9·8, 11·1
Hyoscine	7·6
Hypochlorous acid (T 9·13)	
Hypoxanthine	2·0, 8·9, 12·1
Ibuprofen	4·4, 5·2
Idoxuridine	8·3

Table 9.15 (*contd.*)

Substance	pK_a (approx.)
Imipramine	9·5
Indomethacin	4·5
Inosine	1·5, 8·8
Isoniazid	1·85, 3·5, 10·8
Isoprenaline (isoproterenol)	8·6, 10·1, 12·0
Kanamycin	7·2
Ketamine	7·5
Labetolol	7·4
Lactic acid (T 9·1, footnote)	
Leucine	2·3, 9·7
Levallorphan	4·5, 6·9
Levamisole	8·0
Levodopa	2·3, 8·7, 9·7, 13·4
Levorphanol	8·2
Lignocaine (lidocaine)	7·9
Lincomycin	7·5
Liothyronine (liotrix)	8·5
Lorazepam	1·3, 11·5
Lysergic acid	3·3, 7·8
Lysine	2·2, 9·0. 10·5
Mandelic acid	3·4
Mecamylamine	11·3
Mechlorethamine (chlormethine)	6·4
Meclozine	3·1, 6·2
Mefanamic acid	4·2
Mepacrine (quinacrine, atebrin)	7·9, 10·5
Meperidine (pethidine)	8·7
Mephentermine	10·4
Mepivicaine	7·7
Mepyramine (pyrilamine)	4·0, 8·9
6-Mercaptopurine	7·8, 10·8
Metformin	2·8, 11·5
Methacycline	3·1, 7·6, 9·5
Methadone	8·3
Methaqualone	2·5
Methetoin	8·1
Methicillin	2·8
Methionine	2·2, 9·3
Methotrexate	5·7
Methoxamine	9·2
Methyl 4-hydroxybenzoate	8·4
Methyl nicotinate	3·1
Methylamphetamine	10·1
Methyldopa	2·2, 9·2, 10·6, 12·0

Table 9.15 (*contd.*)

Substance	pK_a (approx.)
Methylene blue	3·8, > 12
Methylergometrine	6·7
Methylphenobarbital	7·8
Methylthiouracil	8·2
Methyprylone	12·0
Methysergide	6·6
Metoclopramide	7·3, 9·0
Metopralol	9·7
Metronidazole	2·5
Miconazole	6·7
Minicycline	2·8, 5·0, 7·8, 9·5
Morphine	8·2, 9·9
Mustine (chlormethine)	6·4
Nafcillin	2·7
Nalidixic acid	6·0
Nalorphine	7·8
Naloxone	7·9
Naphazoline	10·9
Naproxen	4·2
Neostigmine	12·0
Nicotinamide	3·3
Nicotine	3·1, 8·0
Nikethamide	3·5
Nitrazepam	3·2, 10·8
Nitrofurantoin	7·2
Noradrenaline (norepinephrine)	8·6, 9·8, 12·0
Nortriptyline	10·0
Novobiocin	4·2, 9·1
Obidoxime	8·3
Orciprenaline	8·7, 9·9, 11·4
Ornithine	1·7, 8·7, 10·8
Oxazepam	1·7, 11.6
Oxine (T 9·12)	
Oxycodone	8·9
Oxyphenbutazone	4·7
Pamaquin	8·7
Papaverine	6·4
Paracetamol (acetaminophen)	9·5
Parachlorophenol (T 9·4)	
Pargyline	6·9
Pemoline	10·5
Penicillamine	2·4, 7·9, 10·4
Penicillin (see benzylpenicillin)	
Pentazocine	8·5, 10·0

Table 9.15 (*contd.*)

Substance	pK_a (approx.)
Pentobarbital	8·0
Perphenazine	3·7, 7·8
Pethidine (meperidine)	8·7
Phenazocine	8·5
Phenazone (antipyrin)	1·5
Phencyclidine	8·5
Phenethicillin	2·7
Phenformin	2·7, 11·8
Phenindamine	8·3
Pheniramine	4·2, 9·3
Phenmetrazine	8·4
Phenobarbital	7·4
Phenol (T 9·4)	
Phenolphthalein	9·7
Phenoxymethylpenicillin	2·7
Phentermine	10·1
Phentolamine	7·7
Phenylalanine	2·2, 9·3
Phenylbutazone	4·4
Phenylephrine	8·8, 10·1
Phenylmercuric nitrate	3·3
Phenylpropanolamine	9·4
Phenytoin	8·3
Pholcodine	8·0, 9·3
Physostigmine	2·0, 8·1
Pilocarpine	1·6, 6·9
Pimozide	7·3, 8·6
Pindolol	9·7
Piperazine	5·6, 9·8
Polymixin B	8·9
Practolol	9·5
Pralidoxime (2-PAM)	8·1
Prazosin	6·5
Prilocaine	7·9
Probenecid	3·4
Procainamide	9·2
Procaine	9·0
Procarbazine	6·8
Prochlorperazine	8·1
Proflavine (T 9·9)	
Proguanil	2·3, 10·4
Proline	2·0, 10·6
Promazine	9·4

Table 9.15 (*contd.*)

Substance	pK_a (approx.)
Promethazine	9·1
Propicillin	2·7
Propranolol	9·5
Propylthiouracil	8·2
Pseudoephedrine	9·8
Pyrantel	11·0
Pyribenzamine	3·9, 9·0
Pyridoxine	5·0, 9·0
Pyrilamine	4·0, 8·9
Pyrimethamine	7·3
Pyrithione	4·6
Quinalbarbital	7·9
Quinacrine (*see* Mepacrine)	
Quinine	4·1, 8·5
Reserpine	6·6
Resorcinol (T 9·6)	
Retronecine	8·9
Riboflavine	1·9, 10·2
Riboflavine phosphate	2·5, 6·5, 10·3
Rifampicin	1·7, 7·9
Saccharin	1·6
Salbutamol	9·1, 10·4
Salicylic acid (T 9·6)	
Secobutobarbital	8·0
Serine	2.2, 9·2
Serotonin	9·8, 11·1
Sodium cromoglycate	2·5
Sotalol	8·3, 9·8
Spectinomycin	7·0, 8·7
Strychnine	2·3, 8·0
Sulphacetamide	5·4
Sulphadiazine	6·5
Sulphadimethoxine	5·9
Sulphadimidine (sulphamethazine)	7·4
Sulphafurazole	4·9
Sulphamerazine	7·1
Sulphamethizole	5·3
Sulphamethoxazole	5·6
Sulphamethoxypyridazine	7·2
Sulphanilamide	10·4
Sulphapyridine	8·4
Sulphaquinoxaline	5·5
Sulphathiazole	7·1

Table 9.15 (*contd.*)

Substance	pK_a (approx.)
Sulphinpyrazone	2·8
Tartaric acid (T 9·6)	
Terbutaline	8·7, 10·0, 11·0
Tetracaine	8·5
Tetracycline	3·4, 7·8, 9·6
Tetrahydrocannabinol	10·6
Tetramisole	7·8
Thenylpyramine	3·7, 8·9
Theophylline	8·7
Thiamine	4·8, 9·0
Thiamphenicol	7·2
Thiopental	7·6
Thioridazine	9·5
Thiouracil	7·5
Thymidine	9·8, 12.9
Thymine	9·9
Tobramycin	6·7, 8·3, 9·9
Tolazoline	10·6
Tolbutamide	5·3
Tranylcypromine	8·2
Triamterene	6·2
Trichlomethiazide	8·6
Trichloroacetic acid (T 9·1)	
Triethanolamine (T 9·8)	
Trifluoperazine	8·1
Trimethoprim	7·2
Trimeton	9·3
Tripelennamine	3·9, 9·0
Tropicamide	5·2
Tryptamine	10·2
Tryptophan	2·5, 9·5
Tyramine	9·3, 10·9
Tyrosine	2.2, 9·1, 10·1
Uracil	9·4
Uridine	9·3, 12.6
Uridylic acid (UMP)	6·4, 9·5
Urea (T 9·8)	
Valproic acid	5·0
Vinblastine	5·4, 7.4
Vincristine	5·0, 7·4
Viomycin	8·2, 10·3
Warfarin	5·0
Xanthine	7·7, 11·9

(1972), purines and pyrimidines (Albert, 1973), and commonly-prescribed drugs (Martindale, 1982). Regrettably, many results for pharmaceutical products have been published without necessary details such as spread, concentration, or temperature, and this introduces some uncertainty. In spite of this, it is considered that biologically minded readers will find that the listed figures are useful sighting points in charting their own studies.

10 Chelation and the stability constants of metal complexes

The link between this chapter and the preceding ones is the fact that those ionic species which can bind a hydrogen cation have an affinity, also, for the cations of metals. Usually this affinity is greatest when an organic molecule is so designed that it can form at least two bonds to the metal, thus creating a ring. This ring-formation is called *chelation*, from the Greek word for the lobster's claw, and the complexing species is known as a *ligand*. A complexant (chelating agent) may have two or more ionic species, but often only one of these is the ligand.

Metal complexation is a widely discussed topic which brings together physical, organic, and inorganic chemists. Also it is proving of great interest to biochemists, pharmacologists, and molecular biologists. In this chapter, we attempt to give stepwise explanations of the metal-binding process, whose equilibria are expressed by equations similar to those describing ionization. The principal difference is that ionization is commonly regarded as a dissociation, whereas metal binding is seen as an association. These are really nothing more than convenient conventions which lead to stability constants being reported as logarithms (log K), whereas ionization constants are recorded as negative logarithms (pK). The aim in both cases is to obtain convenient whole numbers.

Using the complexation of copper by glycine as our worked example, we proceed in this chapter to show how a stability constant can be conveniently and accurately determined by potentiometric titration, without recourse to a computer program.

Stability constants are the equilibrium constants for reactions between metal ions (M) and ligands (L) which take place in aqueous solution. These equilibria are often represented as:

$$M + nL \rightleftharpoons ML_n \tag{10.1}$$

in which expression the charges on the metal ion, the ligand, and the complex are usually omitted. The hydration of the species are not usually specified. Hence the overall stability constant (β) for equilibrium (10.1) is defined conventionally as

$$\beta_n = \frac{[ML_n]}{[M][L]^n} \qquad (10.2)$$

in which the square brackets represent concentrations in $mol\,l^{-1}$, and β is the product of the microconstants K_1, K_2 etc.

The classic work of Bjerrum (1941) who developed a simple pH-titration method for determining stability constants, created a surge of interest in the quantitative investigation of the many complexes furnished by weak bases (L) or conjugate bases (L^{n-}). At low pH values, such ligands preferentially bind protons rather than metal cations. For example, the overall reaction between a complexant like glycine ($NH_3^+ CH_2 COO^-$) and copper (II) can be represented as:

$$2NH_3^+CH_2COO^- + Cu^{2+} \rightleftharpoons Cu(NH_2CH_2COO)_2 + 2H^+. \qquad (10.3)$$

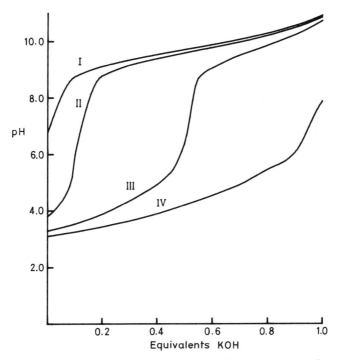

Figure 10.1 Titrations of 0·02M glycine in a medium of 0·15M potassium nitrate at 25 C. Curve I: 0·02M glycine alone, Curve II: 0·02M glycine plus 0·001M copper(II), Curve III: 0·02M glycine plus 0·005M copper(II), Curve IV: 0·02M glycine plus 0·01M copper(II)

Thus, quantitative formation of the CuL_2 complex occurs only if the protons liberated by this equilibrium are removed by reaction with added hydroxyl ions. This is achieved in the pH-titration method by the addition of a standard solution of strong alkali. Titration of the complexant, in the presence of a transition metal ion, results in lower pH values than are observed when the complexant is titrated alone. This can be seen in Fig. 10.1.

The difference between the curves of Fig. 10.1 can, by rather lengthy calculations, reveal the stability constants. A substantial proportion of the many thousands of stability constants recorded in the IUPAC-sponsored compilations by Sillén and Martell (1964, 1971), Perrin (1979), and Högfeldt (1982) have been determined by this pH-titration method, the versatility of which was enhanced considerably, in the 1960s, by the development of computer programs, notably by Sillén (1964) and Sayce (1968). The use of these programs permits many refinements in the calculation of stability constants from the titration data, and allows a greater flexibility in the application of the basic experimental technique. This is because a far wider range of measurements can be made in comparison with what could be contemplated if the calculations were to be performed manually. Moreover, more complicated systems can be investigated. The basic experimental technique, however, remains the same regardless of the type of computational method used to process the data.

It is our belief, that the needs of our readers, for the determination of stability constants which is usually a straightforward process, can be met by the use of a hand-held calculator. Hence we shall confine this chapter to a description of the basic pH-titration method, and the safeguards which must be incorporated when stability constants are calculated manually. Readers requiring information on the computer programs available for calculating stability constants, and how they can be applied, are referred to the book edited by Leggett (1983).

10.1 The nature of chelation

In an aqueous solution of a metal ion, all the coordination positions are usually occupied by water. The reaction shown in equation (10.1) involves replacement of one or more molecules of water by the ligand. When the ligand can release only one pair of electrons to form a coordinate bond, it is said to be *unidentate*, and when two or more pairs can be released, the ligand is said to be *multidentate*. A *chelate* (or chelated complex) is formed when a multidentate ligand forms two or more coordinate bonds with the same metal atom thereby creating the ring structure. Three types of these structures, in which copper (II) and the ligand are present in equimolar amounts, are illustrated in Fig. 10.2.

Note that the charge on the copper cation is not diminished when a 1:1 chelate is formed with ethylenediamine because this ligand contains two electron releasing groups. Glycine, on the other hand, because it has only one electron-releasing group but also has one anionic group, diminishes the charge on the metal by one unit upon chelation with one equivalent, and by two units

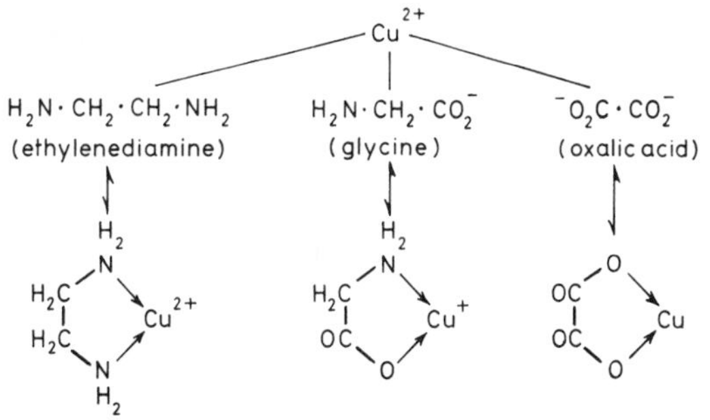

Figure 10.2 Three types of 1 : 1 complex.

when two equivalents of glycine have combined with the metal (see equation 10.3). The third type of chelating structure, exemplified by oxalic acid, bears two anionic groups so that the charge on the metal is decreased by two units after formation of the 1 : 1 complex.

In general, chelation through oxygen or nitrogen takes place only when 5- or 6-membered rings can be formed (because of ring-strain limitations) and of these the 5-membered rings are somewhat more stable. Chelation through sulphur, however, leads to stable 5- and 4-membered rings.

In the presence of an excess of chelating ligand, 2 : 1 complexes can be formed, as described above for copper and glycine. In general, copper(II) has a co-ordination number of four which limits its combinations to a maximum of two ligand molecules. The alkaline earths and manganese(II) behave similarly. Some other cations, notably iron(II), cobalt(II), and zinc(II), have a coordination number of six towards ligands of the ethylenediamine type, and trivalent ions have a coordination number of six towards the glycine type as well.

It must be emphasized that multidentate ligands are not confined to the types which form fairly simple complexes. Multidentate ligands are known which form *polynuclear complexes* by coordinating with two or more central metal atoms. Conversely, a *mixed-ligand* complex can be formed if a single metal atom coordinates with more than one type of ligand (other than water). Often a mixed ligand complex contains both a unidentate ligand and a multidentate chelating ligand, and the mixed complex has different physical and chemical properties from either of the parent complexes. In this chapter, we shall not be concerned with polynuclear or mixed-ligand complexes, but shall confine our treatment to the determination of stability constants of the more usual regular chelates. Yet it is important that the neophyte be alerted to the possibility of competing equilibria, and hence a diagnostic test for such interference will be given (p. 183).

10.2 Method of calculation

Bjerrum (1941) established that complex formation between a unidentate ligand and a metal ion invariably occurs by a stepwise process represented, generically, as:

$$ML_{n-1} + L \rightleftharpoons ML_n \tag{10.4}$$

for which each stepwise formation constant can be defined as:

$$K_n = \frac{[ML_n]}{[ML_{n-1}][L]}. \tag{10.5}$$

The formation of other types of metal chelates was also found to follow this stepwise pattern which, when applied to our example of the copper–glycine reaction, can be represented by the equilibria:

$$Cu^{2+} + NH_2CH_2COO^- \rightleftharpoons Cu(NH_2CH_2COO)^+,$$

and

$$Cu(NH_2CH_2COO)^+ + NH_2CH_2COO^- \rightleftharpoons Cu(NH_2CH_2COO)_2.$$

The corresponding equations for the stepwise formation constants K_1 and K_2 are:

$$K_1 = \frac{[Cu(NH_2CH_2COO)^+]}{[Cu^{2+}][NH_2CH_2COO^-]}$$

and

$$K_2 = \frac{[Cu(NH_2CH_2COO)_2]}{[Cu(NH_2CH_2COO)^+][NH_2CH_2COO^-]}. \tag{10.6}$$

These stepwise formation constants are clearly related to the stability constant (β) for the overall reaction corresponding to (10.1), that is:

$$Cu^{2+} + 2NH_2CH_2COO^- \rightleftharpoons Cu(NH_2CH_2COO)_2$$

by the expression:

$$\beta_2 = K_1K_2 = \frac{[Cu(NH_2CH_2COO)_2]}{[Cu^{2+}][NH_2CH_2COO^-]^2}. \tag{10.7}$$

Returning to the general case:

$$M + nL \rightleftharpoons ML_n,$$

the basic equation for calculating β_n, as defined by equation (10.2), is given by Irving and Rossotti (1953) as:

$$\bar{n} + \sum_1^N (\bar{n} - n)\beta_n [L]^n = 0 \tag{10.8}$$

in which N is the maximum ligand number, and $\bar{n}$ (pronounced en-bar) is the average ligand number which gives the mean number of ligands bound to one metal atom. The term n, which has the same connotation in the above reaction as in equation (10.2), is an integer that has a value $n = 1, \ldots, n = N$. Thus, for systems like the glycine–copper complexes of equations (10.6) and (10.7), in which the maximum ligand number is $N = 2$, equation (10.8) can be written as:

$$\frac{\bar{n}}{(\bar{n} - 1)[L]} = \frac{(2 - \bar{n})[L]}{\bar{n} - 1}\beta_2 - K_1. \tag{10.9}$$

It should be noted that the term β_1 is conventionally replaced in equations of these types by the first stepwise formation constant K_1. The function $[L]$ is the concentration of *free* chelating species which, for a particular point on the titration curve (see Fig. 10.1), denotes the concentration of dianion (if the compound is a diprotic acid like oxalic acid), the concentration of monoanion (if the compound is an ampholyte like glycine), or the concentration of uncharged molecular species (if the compound is a diacidic base like ethylene diamine).

To solve the linear equation (10.9), both $[L]$ and $\bar{n}$ have to be calculated for each point on the titration curve. At this juncture, however, attention must be drawn to the nature of the species involved in the equilibria as typified by equations (10.6) and (10.7), and to the fact that it is not possible to apply activity corrections to these species as was so easily done in the determination of ionization constants, as described in Chapter 3. In the titration of a solution of a chelating agent in the presence of a metal ion, it is clear that there is a much wider diversity of solute species contributing to the ionic strength than is the case when the chelating agent is titrated alone. The calculation of a 'titration-generated' ionic strength, as a first step in the calculation of an activity coefficient, would require that the concentrations of free and bound ligand species be known, together with the charge, distribution, and concentrations of the free and bound metal ions. These calculations would require a value for the formation constant of each complex in the solution, and doubtless an iterative process could be developed for this purpose if reliable values of the constants were available. Frequently, however, the values of the successive formation constants lie closely together, and whilst the value for the overall stability constant (β) can be determined quite reliably, the actual values of its component formation constants are often less certain. It is far more expedient, therefore, to perform measurements of pH upon the solution in which a *constant ionic strength* is maintained throughout the titration, and to convert the derived a_H term to the corresponding c_H term. How this can be done is described on p. 189. In stating and defining the terms in the equations for $[L]$ and $\bar{n}$ (10·10)–(10·12), it will be assumed that this conversion to c_H has been carried out, an operation that allows both of these terms (and hence the values of K_1 and β) to be calculated wholly as concentrations for a particular ionic strength.

The function $[L]$ is calculated from the experimentally derived values of pH, the stoichiometry of the solution, and the protonation constants of the ligand. These protonation constants, usually expressed wholly in terms of concentrations, are the reciprocal of the corresponding (concentration-based) ionization constants, and can be regarded as the stepwise protonation formation constants for the ligand. Thus for the citrate ion ligand (L^{3-}), the first protonation constant

$(K_{H,1})$ refers to the process:

$$L^{3-} + H^+ \rightleftharpoons HL^{2-}$$

for which

$$K_{H,1} = \frac{[HL^{2-}]}{[L^{3-}][H^+]} = \frac{1}{K_{a3}^c}$$

where K_{a3}^c is the third ionization constant expressed in terms of concentrations. It is convenient also to define the overall proton formation constants $\beta_{H,2}$ and $\beta_{H,3}$ as:

$$\beta_{H,2} = \frac{[H_2L^-]}{[L^{3-}][H^+]^2} = \frac{1}{K_{a3}^c K_{a2}^c}$$

$$\beta_{H,3} = \frac{[H_3L]}{[L^{3-}][H^+]^3} = \frac{1}{K_{a3}^c K_{a2}^c K_{a1}^c}$$

For a complexant H_mL that can lose m protons on forming the metal chelate (see p. 179 for further definition), the free ligand concentration $[L]$, when c_H is the derived experimental quantity, is given by:

$$[L] = \frac{mC_{H_mL} - c_{KOH} - c_H + K_w^c/c_H}{c_H K_{H,1} + 2c_H^2 \beta_{H,2} + \cdots mc_H^m \beta_{H,m}} . \tag{10.10}$$

The term K_w^c is the value of the ionic product of water in terms of the concentrations of hydrogen and hydroxyl ions at the ionic strength selected for the determination (see p. 187). It is required only when c_{OH} becomes significantly large in comparison with the other terms of the numerator. At 25°C and an ionic strength of 0·15M with respect to potassium nitrate, the pK_a^c values for glycine were found to be 2·41 and 9·64, yielding:

$$K_{H,1} = \frac{1}{10^{-9\cdot64}} = 4\cdot365 \times 10^9,$$

and

$$\beta_{H,2} = \frac{1}{10^{-9\cdot64}\,10^{-2\cdot41}} = 1\cdot122 \times 10^{12}.$$

The value of pK_w^c at 25°C in a solution of 0·15 M potassium nitrate is estimated, from the data of Jameson and Wilson (1972), to be 13·75. Hence equation (10.10) for this example becomes:

$$[L] = \frac{C_{HL} - c_{KOH} - c_H + 10^{-13\cdot75}/c_H}{4\cdot365 \times 10^9 c_H + 2\cdot244 \times 10^{12} c_H^2} \tag{10.11}$$

The term C_{H_mL} in equation (10.10) is the concentration of protonated complexant present initially, corrected at each point on the titration curve for the volume change caused by the addition of standard alkali.

The function $\bar{n}$ is calculated from $[L]$ using the equation:

$$\bar{n} = \frac{C_{H_mL} - \alpha_{L(H)}[L]}{C_M} \tag{10.12}$$

in which $\alpha_{L(H)} = 1 + c_H K_{H,1} + c_H^2 \beta_{H,2} + \cdots c_H^m \beta_{H,m}$, which is similar in form to the denominator of equation (10.10). The term C_M is the total metal ion concentration initially present in the solution corrected for the added volume of titrant, and C_{H_mL} has already been defined.

For the large number of complexes, in which the total ligand number is two, equation (10.9) can be solved graphically, or by the method of least squares, by plotting values of $\bar{n}/(\bar{n}-1)[L]$ against the corresponding values of $(2-\bar{n})[L]/(\bar{n}-1)$ to yield a straight line of which the slope is β_2 and the intercept is $-K_1$. This can conveniently be carried out by use of a hand-held calculator equipped with a least-squares facility.

It will be noticed, in most instances, that the points corresponding to $\bar{n}$ values from 0·9 to 1·1 are excessively sensitive to experimental error and should be excluded if equation (10.9) is solved by the method of least squares. Generally the most suitable values are obtained for results corresponding to the $\bar{n}$ ranges from 0·2 to 0·8, and from 1·2 to 1·8.

Worked examples of these calculations are given in Table 10.1 for two separate sets of data obtained at identical temperature, ionic strength, and concentrations of complexant (glycine), but at different concentrations of total metal ion, copper (II).

In these examples, the total concentration of glycine taken initially was 0·02M and this was titrated with alkali, in one determination, in the presence of $\sim$ 0·01M copper(II), and in the other determination it was titrated in the presence of $\sim$ 0·005M copper(II). That agreement was found between the values of log K_1 (8·25 and 8·28, respectively) and those of log β_2 (15·14 and 15·16, respectively), for the two sets of results, indicates that any competing equilibria exerted a negligible influence upon this determination.

In general, it is important that at least two determinations be carried out at different complexant:metal ratios because the internal consistency of any one set of results is not evidence for excluding competing equilibria. If a large discrepancy is found, between values of stability constants calculated from two sets of results obtained at different complexant:metal ion ratios, the titrations should first be repeated as a check on technique. Once it has been established that the accuracy of the basic data is satisfactory, it is very likely that competing equilibria are present. It is not possible to deal with these by the calculations described here. Recourse must be made to the more detailed method of calculation afforded by a computer program (see Leggett, 1983). However, it is emphasized that the method of calculation described in this chapter has been found suitable for very many simple chelates, such as are most often encountered.

Table 10.1   Determination of the stability constants of a metal complex

Substance: Glycine $C_2H_5NO_2 = 75.07$. $pK_a^c = 2.41$ and 9.64 in $0.15M$ KNO_3 at $25.0°C$.

Metal: Copper(II) nitrate $0.050\,46M$ prepared in $0.15M$ KNO_3. *Temperature:* $25.0°C$

Glycine $0.02M$ + Copper(II) $0.005M$

$0.15M$ KNO_3 (45.0 ml from burette) + $0.050\,46M$ Cu^{2+} (5.0 ml) purged with N_2 for 15 min after which glycine ($0.075\,07$ g) was added to the solution

$1M$ KOH (ml)	pc_H	C_M	$[L]$	$\bar{n}$	$\dfrac{\bar{n}}{(\bar{n}-1)[L]}$	$\dfrac{(2-\bar{n})[L]}{(\bar{n}-1)}$
0.000	3.217	0.005 05	5.581×10^{-9}	0.578	-2.454×10^8	-1.881×10^{-8}
0.025	3.276	0.005 04	6.446×10^{-9}	0.607	-2.396×10^8	-2.285×10^{-8}
0.050	3.336	0.005 04	7.434×10^{-9}	0.642	-2.412×10^8	-2.820×10^{-8}
0.075	3.398	0.005 04	8.586×10^{-9}	0.681	-2.486×10^8	-3.550×10^{-8}
0.100	3.467	0.005 04	1.007×10^{-8}	0.724	-2.605×10^8	-4.656×10^{-8}
0.150	3.625	0.005 03	1.439×10^{-8}	0.822	-3.209×10^8	-9.523×10^{-8}
0.175	3.709	0.005 03	1.738×10^{-8}	0.866	-3.718×10^8	-1.471×10^{-7}
0.200	3.806	0.005 03	2.141×10^{-8}	0.939		
0.250	4.033	0.005 02	3.499×10^{-8}	1.077		
0.300	4.279	0.005 02	5.882×10^{-8}	1.233	8.997×10^7	1.936×10^{-7}
0.325	4.417	0.005 01	7.847×10^{-8}	1.323	5.220×10^7	1.645×10^{-7}
0.350	4.556	0.005 01	1.047×10^{-7}	1.410	3.285×10^7	1.507×10^{-7}
0.375	4.698	0.005 01	1.402×10^{-7}	1.502	2.134×10^7	1.391×10^{-7}
0.400	4.858	0.005 01	1.950×10^{-7}	1.596	1.373×10^7	1.322×10^{-7}
0.450	5.275	0.005 00	4.691×10^{-7}	1.787	4.840×10^6	1.270×10^{-7}
0.500	6.345					

Intercept $= -1.893 \times 10^8$ Slope $= 1.457 \times 10^{15}$

$\log K_1 = 8.28$ $\log \beta_2 = 15.16$

10.3 Choice of ionic medium and the preparation of standard solutions

The choice of a salt from which a medium of constant ionic strength can be prepared, for the determination of stability constants, is governed by four factors:

1. The salt must be very soluble in water;
2. The complexation between its anion and the metal cation used in the determination, must be negligible;
3. There must be no reaction between the salt and the solution used in the salt bridge fitted to the reference half cell; and
4. It should be a salt for which the concentration ionic product of water

Table 10.1 (*contd.*)

Glycine 0·02M + Copper(II) 0·01M
0·15M KNO_3 (40·0 ml from burette) + 0·050 46M Cu^{2+} (10·0 ml) purged with N_2 for
15 min after which glycine (0·075 07 g) was added to the solution

1M KOH (ml)	pc_H	C_M	$[L]$	$\bar{n}$	$\dfrac{\bar{n}}{(\bar{n}-1)[L]}$	$\dfrac{(2-\bar{n})[L]}{(\bar{n}-1)}$
0·0	3·027	0·010 09	$3·133 \times 10^{-9}$	0·401	$-2·137 \times 10^{8}$	$-8·363 \times 10^{-9}$
0·05	3·105	0·010 08	$3·781 \times 10^{-9}$	0·437	$-2·053 \times 10^{8}$	$-1·050 \times 10^{-8}$
0·10	3·182	0·010 07	$4·504 \times 10^{-9}$	0·481	$-2·058 \times 10^{8}$	$-1·318 \times 10^{-8}$
0·15	3·269	0·010 06	$5·471 \times 10^{-9}$	0·528	$-2·045 \times 10^{8}$	$-1·706 \times 10^{-8}$
0·20	3·359	0·010 05	$6·626 \times 10^{-9}$	0·581	$-2·093 \times 10^{8}$	$-2·244 \times 10^{-8}$
0·25	3·468	0·010 04	$8·368 \times 10^{-9}$	0·635	$-2·079 \times 10^{8}$	$-3·129 \times 10^{-8}$
0·30	3·567	0·010 03	$1·013 \times 10^{-8}$	0·704	$-2·348 \times 10^{8}$	$-4·435 \times 10^{-8}$
0·35	3·687	0·010 02	$1·280 \times 10^{-8}$	0·775	$-2·691 \times 10^{8}$	$-6·969 \times 10^{-8}$
0·40	3·817	0·010 01	$1·638 \times 10^{-8}$	0·851	$-3·487 \times 10^{8}$	$-1·263 \times 10^{-7}$
0·50	4·126	0·009 99	$2·899 \times 10^{-8}$	1·016		
0·60	4·470	0·009 97	$5·230 \times 10^{-8}$	1·199	$1·152 \times 10^{8}$	$2·105 \times 10^{-7}$
0·65	4·659	0·009 96	$7·118 \times 10^{-8}$	1·294	$6·183 \times 10^{7}$	$1·709 \times 10^{-7}$
0·70	4·856	0·009 95	$9·633 \times 10^{-8}$	1·391	$3·693 \times 10^{7}$	$1·500 \times 10^{-7}$
0·75	5·117	0·009 94	$1·470 \times 10^{-7}$	1·488	$2·074 \times 10^{7}$	$1·542 \times 10^{-7}$
0·80	5·388	0·009 93	$2·201 \times 10^{-7}$	1·586	$1·230 \times 10^{7}$	$1·555 \times 10^{-7}$
0·85	5·609	0·009 92	$2·743 \times 10^{-7}$	1·686	$8·960 \times 10^{6}$	$1·256 \times 10^{-7}$
0·90	5·987	0·009 91	$4·378 \times 10^{-7}$	1·784	$5·198 \times 10^{6}$	$1·206 \times 10^{-7}$
1·00	7·797					

Intercept $= -1·791 \times 10^{8}$ Slope $= 1·381 \times 10^{15}$
$\log K_1 = 8·25$ $\log \beta_2 = 15·14$

(K_w^c) is known at the concentration of salt selected and the temperature of the determination.

Sodium perchlorate, at a concentration of 3·0M, has often been used as a medium of constant ionic strength for these determinations because the perchlorate ion shows little tendency to form complexes with transition metal ions. A disadvantage of this solution, however, is that the usual calomel reference half cell cannot be employed directly in the cell used for the measurement of pH because precipitation of potassium perchlorate in the junction of the half cell (from KCl) causes large and unstable junction potentials. Ideally the same ionic medium should be used on both sides of the salt bridge and this ideal is approached for test solutions prepared in 3·0M sodium perchlorate by the use of the following reference half cell (Rossotti and Rossotti, 1961):

$$Ag|AgClO_4(0.010M), NaClO_4(2.990M)|NaClO_4(3.000M)\ bridge|test\ soln.$$

$$(10.13)$$

The other most often employed salt is potassium nitrate, which we prefer. We recommend the concentration of 0.15M because, in the presence of both complexant and the metal ion (at the concentrations usually employed for determinations), the total ionic strength approximates to the physiological value of $\sim 0.16M$. This medium offers the advantage that the calomel reference half cell can be used without modification. It is assumed that the diffusion of chloride ion through the salt bridge is so small that the resulting formation of metal–chloro complexes will be negligible in the presence of a comparatively enormous excess of chelating agent during the actual determination. Nonetheless, the disparity between the concentrations and properties of the 0.15M potassium nitrate solution on one side of the junction and the reference half cell solution of 3.5M, 3.8M, or *saturated* potassium chloride on the other side can be expected to yield a greater liquid junction potential than would be the case for the reference half cell specified in (10.13).

Fortunately, compensation for this disparity between the two half cells is inherent in the method described on p. 189 for the experimental determination of the factor for converting a measured hydrogen ion activity to the corresponding hydrogen ion concentration. In addition to the activity coefficient of the hydrogen ion, this factor contains a contribution for the liquid junction potential. Provided that the same electrodes are used in the determination of the pK_a^c values and of the stability constants, the application of the factor for the conversion of the measured pH to pc_H can be assumed to compensate also for the liquid junction potential.

In the list of four points to be considered when choosing the background salt solution, the fourth point concerned the advisability of knowing the value of K_w^c for this solution at the appropriate temperature(s). This value is rarely needed in calculating $[L]$ by equation (10.10), because the pH values obtained during the titration of the complexant in the presence of metal cation are such that c_{OH} makes a negligible contribution to the numerator of this equation. In many cases, however, when the pK_a value of the ligand species is being calculated, allowance must be made for the hydroxyl ion concentration, using the type of method described in Chapter 2. However, it is not valid, in solutions of comparatively high ionic strengths, to make the assumption implicit in the equations of Chapter 2 that the thermodynamic value of the ionic product of pure water (K_w) suffices for the calculation of the hydroxyl ion concentration. For example, the calculation of c_{OH} from the derived value of pc_H obtained in 0.15 M potassium nitrate requires that the value of the concentration ionic product (K_w^c) be known, because this quantity is defined as:

$$K_w^c = c_H c_{OH}. \qquad (10.14)$$

Table 10.2 Values of pK_w^c for potassium nitrate
solutions at various temperatures

Conc.	pK_w^c			
M	15°C	25°C	35°C	45°C
0·06	14·159	13·809	13·491	13·203
0·10	14·132	13·778	13·457	13·168
0·14	14·116	13·753	13·441	13·154
0·18	14·101	13.749	13·423	13·138

Values of pK_w^c (i.e. $-\log K_w^c$) for potassium nitrate solutions, as obtained by Jameson and Wilson (1972), are given in Table 10.2. In the example of Table 10.1, we used the value 13·75 (at 25°C) for calculating $pK_a^c = 9·64 \pm 0·01$ of the glycine ligand at a concentration of 0·02M in 0·15M potassium nitrate.

Now that the factors which govern the choice of the medium for creating a constant ionic strength have been discussed, the use of 0·15M potassium nitrate is to be assumed throughout the rest of this chapter. It is important that this solution be used for dissolving *all* substances used in the determination of the ionization constants and stability constants. We found it expedient to prepare an approximately 0·15M solution from an accurately weighed, but undried, portion of analytical-reagent grade potassium nitrate. A known aliquot of this solution (20·0 ml) was passed through a prepared column of Amberlite IR 120H resin ($\sim$ 50 ml) which was then washed with water (3 × 50 ml portions). The eluate and washings were collected in a 250-ml beaker and the concentration of acid produced by the ion exchange was determined by pH titration using standardized 0·1M alkali. From the results, the exact weight required to make 1 litre of 0·015M potassium nitrate was calculated and this amount was dissolved in water and diluted to 1 litre (solution 1). After checking that the concentration of this solution was, in fact, 0·15M, the same weight of salt was transferred separately into each of four other 1-litre graduated flasks (the solid dissolved in the amount of water required to effect this transfer). Standard solutions of 1M potassium hydroxide, 0·1M nitric acid, 1M hydrochloric acid, and copper(II) nitrate were eventually prepared in these flasks. For the 1M potassium hydroxide and 1M hydrochloric acid solutions, we used the commercially available ampoules which, on transfer and dilution of the contents, yielded 1·000M solutions. In the case of the 1M potassium hydroxide prepared in 0·15M potassium nitrate solution, the water used for dilution was purged of carbon dioxide by passing nitrogen gas through it for about 15 min. The nitric acid solution was prepared by adding a slight excess of the calculated quantity of concentrated nitric acid to the 1-litre graduated flask containing potassium nitrate, and diluting the solution to the mark with water. The solution was standardized against alkali and any adjustment of the volume, as required to make the solution 0·100M, was made with the original 0·15M solution of potassium nitrate (solution 1, above). The molarity of this solution of nitric acid was checked by pH titration. The 0·05M copper(II) nitrate solution was prepared by adding a slight excess of the calculated weight of analytical-reagent grade solid to a flask containing the weighed portion of potassium nitrate, and diluting the solution to 1 litre with water. The concentration of copper(II) in this solution was determined electrogravimetrically as copper metal, and the volume of the solution was adjusted with solution 1 so that the copper(II) concentration was 0·05M. This concentration was checked by electrogravimetric determination.

10.4 Measurement of pH and calculation of pc_H

The apparatus used for the measurement of pH and for the titrations was similar in type to that described in Chapter 3, except that in this instance the pH meter was calibrated in 0·01 pH unit. The output from this instrument was also connected to a pen recorder in order to establish the equilibrium value of pH. Such an arrangement is a valuable aid, particularly in the calibration procedure which, in this instance, involved buffers with pH values of 6·865 (0·025 m KH_2PO_4, 0·025 m Na_2HPO_4), 4·008 (0·05 m KH phthalate), and 9·180 (0·01 m borax), at 25°C. On transferring the *electrodes* between these buffer solutions, which were contained in stoppered tall-form beakers positioned in a thermostatically controlled water bath, each electrode was washed three times with the buffer solution into which it was about to be immersed. However, when they were transferred into the glycine solutions prepared in 0·15M potassium nitrate, they were washed with 0·015M potassium nitrate solution contained in a wash bottle that was normally stored in the water bath at 25°C.

The glycine solutions were contained in a beaker sealed into a jacket through which water from the bath was circulated, and this beaker was mounted on the platform of a magnetic stirrer. The solutions were prepared in the beaker by adding, from a burette, the required volume of 0·15M potassium nitrate followed, if appropriate, by the copper(II) solution transferred to the beaker by pipette to ensure the volume was 50·0 ml before the titration. Nitrogen gas, saturated with respect to water vapour by passage through two Dreschel bottles each containing 0·15M potassium nitrate, was bubbled through the solution in the jacketted beaker for a minimum of 15 min in order to remove dissolved carbon dioxide and oxygen. Glycine (0·07507 g, dried at 120°C for 1 hour) was then transferred to the solution from a tared watchglass by means of a camel-hair brush, and dissolved while the solution was being stirred. The titrant was added from an Agla Micrometer Syringe through a length of cannula tubing attached to the luer-lock fitting needle, and the solution was stirred during the course of each addition. The tip of this cannula tube was removed from the solution immediately after each addition. Both the stirring of the solution and the flow of nitrogen through it were stopped while the pH was being monitored by the

Table 10·3 Determination of the factor for the conversion of the measured pH to pc_H at 25°C

0·15M Potassium nitrate (45·0 ml from burette) purged with nitrogen for 15 min. 0·100M nitric acid in 0·15M potassium nitrate added, and the solution titrated with 1·000M potassium hydroxide

1M KOH	pH	C_t	c_{KOH}	c_{HNO_3}	pc_H*	pH − pc_H
0·0	2·073	0·010 00		0·010 00	2·000	0·073
0·05	2·120	0·010 00	0·001 00	0·009 00	2·046	0·074
0·10	2·170	0·009 98	0·002 00	0·007 98	2·098	0·072
0·15	2·229	0·009 97	0·002 99	0·006 98	2·156	0·073
0·20	2·298	0·009 96	0·003 98	0·005 98	2·223	0·075
0·25	2·373	0·009 95	0·004 98	0·004 97	2·304	0·069
0·30	2·471	0·009 94	0·005 96	0·003 98	2·400	0·071
0·35	2·597	0·009 93	0·006 95	0·002 98	2·526	0·071
0·40	2·771	0·009 92	0·007 94	0·001 98	2·703	0·068
0·45	3·092	0·009 91	0·008 92	0·000 99	3·004	0·088

*$pc_H = - \log c_{HNO_3}$.
Conversion factor $= 0.073 \pm 0.015$ (Standard deviation 0·006).

chart recorder. The pH was measured when the recording indicated that an equilibrium value had been attained.

The factor required for the conversion of the measured values of pH into the corresponding pc_H values was determined by two methods. In the first, duplicate titrations of solutions (50·0 ml) of 0·100M nitric acid, (prepared in 0·15M potassium nitrate) with 1·000M potassium hydroxide that had also been prepared in 0·15M potassium nitrate, were performed. The measured value of pH was recorded after each addition as is shown in the specimen determination given in Table 10.3. The difference between the pc_H value, calculated from the stoichiometric concentrations of the solution and the measured pH value, was taken as the conversion factor.

In the second method, duplicate titrations of solutions of 0·15M potassium nitrate (50·0 ml) with 1·000M hydrochloric acid were similarly carried out. The four values of the conversion factor obtained were:

$$HNO_3 \qquad HCl \qquad Mean$$
Factor (pc_H − pH)   0·062   0·073   0·071   0·084   0·073 ± 0·011.

The third decimal place has no real significance in these values because the pH meter was calibrated in 0·01 pH divisions. However, the change in the pH values of the standardizing buffers recorded before and after the determinations did not exceed 0·01 pH unit and were estimated as being often within 0·005 pH unit of the original values. In view of these limitations, the results obtained for the conversion factors are in fair agreement. The mean of the values, 0·073, was taken as the factor which had to be subtracted subsequently from all measured values of pH obtained in the glycine titrations to convert them into estimates of pc_H. The latter are the values recorded, for example, in Table 10.1.

It may be asked whether the conversion of the measured pH to pc_H may introduce any error additional to that which is inherent in the types of pH measurements discussed in Chapter 3. However, reference to Table 10.3 will confirm that the range of pH values covered in the determination of the conversion factor is small in comparison with the range of pH covered by the titrations of glycine. Thus we have assumed that the numerical value of the factor remains constant over the whole range of pH covered firstly by the determination of the two pK_a values for glycine (pH $\sim 2\cdot3-10\cdot9$), and secondly in the range covered by the titrations represented in Table 10.1. There is good reason for supposing that this assumption is valid because the results of Avdeef and Bucher (1978), also work quoted by them, confirm that over a wide range of pH the relation between the measured pH and pc_H takes the linear form:

$$pH = A + Spc_H. \qquad (10.15)$$

At an ionic strength of $0\cdot15$ for both potassium and sodium chloride solutions, the value of the slope (S) is close to unity, and under this condition the term A equates to what we have called the conversion factor. It is also worth recalling that this factor is a composite of functions of the activity coefficient of the hydrogen ion and the liquid junction potential. The constancy of this potential is assumed in all pH measurements; also in the transfer of the electrodes from a standardizing buffer solution into a solution of unknown pH, the value of the liquid junction potential is expected to remain invariant.

The liquid junction potential, however, does vary with the ionic strength of the solution, and the more the properties of the solution under test resemble those of the reference half cell solution, the less is the effect of the contribution of this potential. It is likely, therefore, that the error from this cause in solutions of glycine prepared in $0\cdot15M$ potassium nitrate is less than that of similar solutions prepared in water. The overall result is that the error produced by the conversion of pH to pc_H is not likely to be greater than the uncertainty of $0\cdot02$ pH unit inherent in all measurements of pH (see Appendix VI).

10.5 Common difficulties and how they can be overcome

Errors in the determination of stability constants differ little from those observed in other types of pH titrations. The same care must be exercised, for example, in the standardization of the pH meter as is customary for the determination of pK_a values. All substances used in the determinations must be of high purity and their solutions must be accurately standardized. An additional possible source of error, not present in the determination of pK_a values by titration, is the precipitation of a metal hydroxide, or of an insoluble complex at some stage during the titration. A rapid drift to lower pH is often symptomatic of this occurrence and the titration should be stopped because it will no longer be possible to obtain concordant results. The titration can be repeated at a lower metal ion concentration: if this is done, the incremental values of titrant added should also be decreased, as follows from the results of Table 10.1 and expressed graphically in Fig. 10.1. It is possible to determine the stability constants of some chelates in very dilute solution, such as 7·5 μM (Albert and Serjeant, 1960).

The value of m to use in equation (10.10) is not always obvious because, when $m = 2$, both protons must be liberated in forming the 1:1 complex. That this is not the case for glycine–copper(II), is obvious from curve III of Fig. 10·1, because an equivalence point is seen when about 0·5 mmol potassium hydroxide has been added to the 1·0 mmol glycine (i.e. 50·0 ml, 0·02M) taken originally. This quantity of alkali is equivalent to the 2 mole equivalents of protons liberated by the reaction between 0·25 mmol copper(II) and 0·50 mmol glycine, as implied by reaction (10·3). The other 0·5 mmol potassium hydroxide is used in the neutralization of the remaining 0·5 mmol glycine and, if necessary, the proton-lost pK_a value of glycine could be calculated from these results. If, therefore, in the titration curve obtained when the complexant and the divalent metal ion are present in the stoichiometric mole ratio 4:1, there is no equivalence point when about 0·5 mol equivalent of alkali has been added, but good evidence of one when 1·0 equivalent has been added: then, it can be concluded that 4 mol protons are liberated on forming the 2:1 complex. The value m to use in equation 10.10, for this case, would be 2, which indicates that the behaviour of the complexant is consistent with the formula H_2L.

11   Appendices

APPENDIX I

An Outline of the Brønsted–Lowry Theory
(See also p. 2)

The theory proposed by Brønsted (1923) defines an acid as any substance that can ionize in solution to give a solvated hydrogen ion (i.e. a proton stabilized by interaction with either the solvent or a substance in solution). Conversely a base is a substance which can accept a hydrogen ion. Thus an acid is a 'proton donor' and a base a 'proton acceptor' and the ionization process always involves the two which are known as 'conjugate acid–base pairs'. This concept may be generalized by the equation

$$\text{Acid}_1 + \text{Base}_2 \rightleftharpoons \text{Acid}_2 + \text{Base}_1 \qquad (11.1)$$

in which the proton is partitioned between the two bases and the equilibrium is determined by the relative strengths of the acids. $Acid_1$ and $Base_1$ bear a conjugate relation to one another, and so do $Acid_2$ and $Base_2$. As an illustration of the interaction between an acid HA and water we can write

$$HA + H_2O \rightleftharpoons H_3O^+ + A^- \tag{11.2}$$

i.e. $$Acid_1 + Base_2 \rightleftharpoons Acid_2 + Base_1.$$

Two specific examples are

$$CH_3COOH + H_2O \rightleftharpoons H_3O^+ + CH_3COO^- \tag{11.2a}$$

$$NH_4^+ + H_2O \rightleftharpoons H_3O^+ + NH_3. \tag{11.2b}$$

In equation (11.2a) the acetate ion (*the anion*) is the conjugate base of acetic acid (*the neutral species*), and in equation (11.2b) the ammonium ion (*the cation*) is the conjugate acid of the base ammonia (*neutral species*). In both cases, water has acted as a base because it has accepted a proton from the acidic species (CH_3COOH and NH_4^+, respectively) and stabilized it. It should be emphasized, however, that the exact degree of hydration of the hydrated proton is not known, and the monohydrated species H_3O^+ is invoked merely to illustrate the essential role of solvent in the ionization process. In the pH range 0–14 strong acids are completely ionized and for them equation (11.2) becomes stoichiometric (i.e. $HA + H_2O \rightarrow H_3O^+ + A^-$. Weak and moderately strong acids, on the other hand, are incompletely ionized, although ionized more completely in dilute than in concentrated solutions. All the species shown in equation (11.2) are coexistent to some degree; but in solutions of the stronger acids, the species on the right hand side of the equation are more favoured than in solutions of weaker acids.

Anions derived from the weaker acids undergo the reaction with water known as hydrolysis. This process is also described by equation (11.1). For example if sodium acetate, an ionic solid, is dissolved in water, hydrolysis of the acetate ion occurs.

$$CH_3COO^- + H_2O \rightleftharpoons CH_3COOH + OH^- \tag{11.3}$$

i.e. $$Base_1 + Acid_2 \rightleftharpoons Acid_1 + Base_2.$$

In this case water has acted as a proton donor. It acts similarly when weak or moderately strong bases are dissolved. For example, ammonia solutions contain equimolar (but small) amounts of the ammonium ion and the hydroxyl ion

$$NH_3 + H_2O \rightleftharpoons NH_4^+ + OH^-$$

and, in general

$$B + H_2O \rightleftharpoons BH^+ + OH^- \tag{11.4}$$

where B represents any weak or moderately strong base and BH^+ is the cation.

The Brønsted–Lowry theory and earlier theories of ionization* applied the *law of mass action*, subsequently confirmed thermodynamically, to derive an ionization constant which represents the state of the equilibrium. In its simplest form, the ratio of the products (of the concentrations of species produced by the ionization) always bears a fixed ratio to the concentration of the species from which they were derived. Thus the products of the concentration of *ions* derived from acetic acid always bears a fixed ratio to the concentration of non-ionized *molecules*. For ionizations in dilute aqueous solution, it is customary to neglect the role of water because its concentration is not altered by the ionization. The fixed ratio referred to above is taken, therefore, as applying to $[H_3O^+][CH_3COO^-]/[CH_3COOH]$. This ratio is called the acidic ionization constant (K_a) or more simply the ionization constant. Thus

$$K_a = \frac{[H_3O^+][CH_3COO^-]}{[CH_3COOH]} \tag{11.5}$$

describes the equilibrium $CH_3COOH + H_2O \rightleftharpoons CH_3COO^- + H_3O^+$ and this has been found by experiment to be $1{\cdot}75 \times 10^{-5}$ mol l^{-1} at $20°C$. For the sake of brevity 'mol l^{-1}' is usually omitted and, in this book, we have abbreviated the ionization of acids to the generalization

$$HA \rightleftharpoons H^+ + A^- \tag{11.6}$$

which yields the equation

$$K_a = \frac{[H^+][A^-]}{[HA]} \tag{11.7}$$

whence

$$pK_a = pH + \log[HA] - \log[A^-]. \tag{11.7a}$$

Where A^- is any anion and p means the negative logarithm, whence acetic acid has pK_a 4.76.

The ionization of conjugate acids is described similarly, thereby providing a convenient scale on which to assess the strength of bases. Equation (11.2b) (p. 193), for example, describes the ionization of the ammonium ion, the *cation* of the base ammonia. To make this equation analogous to equation (11.6) we can write

$$NH_4^+ \rightleftharpoons NH_3 + H^+$$

and in general:

$$BH^+ \rightleftharpoons B + H^+. \tag{11.8}$$

---

*Including the first theory of ionization worked out by S. Arrhenius between 1884 and 1887 (see Bell, 1969).

Thus the ionization constant for any given base is expressed as

$$K_a = \frac{[B][H^+]}{[BH^+]} \tag{11.8a}$$

whence

$$pK_a = pH - \log[B] + \log[BH^+]. \tag{11.8b}$$

For the ammonium ion this has been found by experiment to be 9·2 at 25°C.

K_b values

In the older literature the ionization constants of bases referred to the equilibrium described by equation (11.4). These constants were known as K_b. Thus for ammonia

$$K_b = \frac{[NH_4^+][OH^-]}{[NH_3]}$$

and K_b was found experimentally to be $1·8 \times 10^{-5}$ at 25°C (i.e. pK_b 4·74). These constants are related to pK_a values by equation (1.7) on p. 6. Today the use of K_b is avoided because it does not touch the heart of the matter, namely: an acid produces hydrogen ions and a base receives them. Thus both acids and bases are best related in terms of a single quantity, their affinity for the hydrogen

Table 11.1   Interconversion of K_a and pK_a

To convert K_a to pK_a, take the logarithm of the constant and subtract this from zero. For example, acetic acid has the ionization constant $1·75 \times 10^{-5}$; the logarithm $-4·7570$, subtracted from 0, gives 4·7570, which is the pK_a. Bases, when expressed as pK_a, are converted in exactly the same way.

   To convert pK_a to K_a, simply use the equation: $K_a = 10^{-pK_a}$.
Practice may be had with the following examples:

	$pK_a(20°C)$	K_a
Benzoic acid	4·12	$7·6 \times 10^{-5}$
p-Nitroaniline	1·01	$9·8 \times 10^{-2}$
Ammonia	9·36	$4·4 \times 10^{-10}$
p-Cresol	10·14	$7·2 \times 10^{-11}$
Methylamine	10·81	$1·5 \times 10^{-11}$

   When bases are expressed as K_b, as in the older literature, it is convenient to convert them first to pK_b (using the same procedure as for pK_a) and then to convert to pK_a by subtracting the result from 14·17 (at 20°C), or 14·00 (at 25°C). For example, one reported K_b value for ammonia is $1·5 \times 10^{-5}$ (at 20°C). The logarithm of which is $-4·8239$. Subtracting 4·82 from 14·17 gives 9·35 which is the pK_a at 20°C.

ion. Such a relationship requires the use of the acidic constant (K_a) for *both* acids and bases.

Table 11.1 provides elementary assistance in the interconversion of K_a and pK_a values.

APPENDIX II

Comparison of classical and thermodynamic quantities

At a given temperature, a true ionization constant is a thermodynamic quantity related to the standard free energy change in the reaction. Theoretically this value should be independent of the concentration taken initially for its determination. In practice, however, application of equation (11.7) to the determination of the ionization constant of acetic acid yields values of K_a which vary with concentration as shown in Table 11.2.

These and similar variations observed for other acids and bases arise principally from electrostatic interaction of the ions produced. This interaction becomes greater with higher concentrations, hence some of each ionic species in equation (11.7) is not completely free. The molar concentrations of these ions, as indicated by the square brackets, are, therefore, greater than the concentrations at which the ions are free and active in the solution. Although these active concentrations are somewhat less than the stoichiometric molar concentrations in dilute solutions, both figures become increasingly convergent as the total concentration is decreased. The K_a of equation (11.7) is therefore called the concentration ionization constant (K_a^c) because it varies with the dilution*. The true thermodynamic constant (K_a^T) must be expressed in terms of active concentrations and these are commonly termed *activities*. Thus we

Table 11.2 Concentration ionization constants of acetic acid at 25°C

Total concentration CH_3COOH (mol l^{-1})	$K_a \times 10^5$	pK_a
0·000 03	1·768	4·752 5
0·000 22	1·781	4·749 4
0·002 4	1·809	4·742 4
0·009 8	1·832	4·737 0
0·020 0	1·840	4·735 2

*This concentration constant is sometimes called 'the classical constant' and it is related to the 'apparent constant' (K'); but these are less desirable terms.

should correctly write

$$K_a^T = \frac{a_{H^+} \cdot a_{A^-}}{a_{HA}}$$

more conveniently written as

$$K_a^T = \frac{\{H^+\}\{A^-\}}{\{HA\}} \qquad (11.9)$$

where a_{H^+}, a_{A^-}, and a_{HA} refer to the activities of the three species involved. At infinite dilution the concentration constant becomes numerically equal to the thermodynamic constant. The extrapolated value from the concentration ionization constants of acetic acid given in Table 11.2 is 1.752×10^{-5} at infinite dilution and this is the K_a^T. For *univalent* acids and bases, the difference between K_a^c and K_a^T is small in 0·01 M, and usually negligible in 0·001 M, concentrations. Equation (11.7) can often be used for the sake of simplicity provided that constants are determined in solutions not stronger than 0·01 M and only *univalent* ions are present. The conditions under which allowance for activity effects *must* be made are discussed in Chapter 3.

APPENDIX III

Calculations of hydrogen ion activity and concentration; also of hydroxyl ion activity and concentration

In potentiometric work, it is often required to tabulate the hydrogen ion *activity* $\{H^+\}$, obtainable as follows:

$$\{H^+\} = 10^{-pH}.$$

Thus, for the $\{H^+\}$ equivalent to pH 3·25,

$$\{H^+\} = 10^{-3.25}$$
$$= 0.000\ 562\ 3.$$

In other cases (e.g. Table 2.9), the hydroxyl ion *activity* $\{OH^-\}$ must be calculated from the pH.

From the equation $pOH^- = pK_w - pH$, we obtain:

$$\{OH^-\} = 10^{(pH - pK_w)}.$$

(Values of pK_w at various temperatures are in Table 11.8.)

Example
Calculate $\{OH^-\}$ equivalent to pH 10·71 at 20°C

$$\{OH^-\} = 10^{(10.71 - 14.17)}$$
$$= 0.000\ 35.$$

Table 11.3   Correlation of pH, $\{H^+\}$ and $\{OH^-\}$

pH	$\{H^+\}$	pH	$\{OH^-\}^{20\,C}$	$\{OH^-\}^{25\,C}$
4·0	0·000 100 0	10·0	0·000 068	0·000 10
3·9	0·000 125 9	10·1	0·000 085	0·000 13
3·8	0·000 158 5	10·2	0·000 11	0·000 16
3·7	0·000 199 5	10·3	0·000 135	0·000 20
3·6	0·000 251 2	10·4	0·000 17	0·000 25
3·5	0·000 316 2	10·5	0·000 21	0·000 32
3·4	0·000 398 1	10·6	0·000 27	0·000 40
3·3	0·000 501 2	10·7	0·000 34	0·000 50
3·2	0·000 631 0	10·8	0·000 43	0·000 63
3·1	0·000 794 3	10·9	0·000 54	0·000 79
3·0	0·001 000	11·0	0·000 68	0·001 0
2·3	0·005 012	11·7	0·003 4	0·005 0
2·0	0·010 00	12·0	0·006 8	0·010
1·3	0·050 12	12·7	0·034	0·050
1·0	0·100 0	13·0	0·068	0·10

The figures in Table 11.3 are intended as sighting values and for checking the position of decimal point.

In spectrometric work, solutions of standard hydrogen ion *concentration* are often required. These are obtained by diluting hydrochloric acid to the required concentration, because complete ionization can be assumed. In Table 11.4, examples of $p[H^+]$ obtained in this way are compared with the corresponding hydrogen ion activity, written as $p\{H^+\}$ (as measured potentiometrically and usually referred to simply as pH). These results were obtained at 20°C. The temperature correction is only $+0.0003°C^{-1}$.

Solutions of standard hydroxyl ion *concentration* are also often required in spectrometric work, and are usually obtained with dilutions of sodium hydroxide. In Table 11.5, examples of $p[H^+]$ obtained in this way are compared with the corresponding hydrogen ion activity $p\{H^+\}$ as calculated from y (the ionic activity coefficients). These data are presented for 25°C, at which

Table 11.4   Solutions of standard hydrogen ion concentration

HCl N	$[H^+]$	$p[H^+]$	$p\{H^+\} = pH$
0·001	0·001	3·00	3·02
0·01	0·01	2·00	2·05
0·02	0·02	1·70	1·77
0·05	0·05	1·30	1·41

Table 11.5   Solutions of standard hydroxyl ion concentration

1	2	3	4	5	6
NaOH M ($25°C$)	$[OH^-]$	$p[OH^-]$	$p[H^+]$	log y^{NaOH}	$p\{H^+\}$ = pH (sum of columns 4 and 5)
0·01	0·01	2·00	12·00	$\bar{1}$·95	11·95
0·02	0·02	1·70	12·30	$\bar{1}$·93	12·23
0·05	etc.	1·30	12·70	$\bar{1}$·91	12·61
0·10		1·00	13·00	$\bar{1}$·88	12·88
0·20		0·70	13·30	$\bar{1}$·86	13·16
0·50		0·30	13·70	$\bar{1}$·83	13·53
1·00		0·00	14·00	$\bar{1}$·82	13·82
2·00		− 0·30	14·30	$\bar{1}$·84	14·14

most of the figures have been obtained.* The temperature correction is large, namely $− 0.0355$ unit $°C^{-1}$. Thus, the pH of 0·01N NaOH is 12·84 at $0°C$, 12·13 at $20°C$ and 11·50 at $38°C$.

APPENDIX IV

Some effects of temperature on ionization constants

In general, organic bases have much greater temperature coefficients than organic acids. A comparison of calculated and observed values of the temperature coefficient, for a selection of nitrogenous bases, is given in Table 11.6 as an indication of the validity of equation (1.8) (see p. 12), which is suitable for the range 15–35°C.

Outside this temperature range (288 – 308 K) (because the standard entropy change, $\Delta S°$, is temperature-dependent), the more rigorous equation

$$\frac{- d(pK_a)}{dT} = \frac{pK_a + 0.218 \Delta S^0}{T}$$
(11.10)

must be used.

Bases which have *two* ionizable groups (diacidic bases) require a different equation (11.11) to describe the temperature dependence of the pK_a for the process $H_2B^{2+} \rightleftharpoons H^+ + BH^+$. This represents the ionization process of the *dication*, namely the weaker of the two groups, and hence the lower pK_a value. The temperature–pK_a relationship for the monocation, the stronger of the

*Harned and Hecker (1933).

Table 11.6   Temperature coefficients for univalent organic cations

Substance	pK_a at 25°C	$-d(pK_a)/dT$	
		Calc.	Found
2-nitroaniline	-0.26	-0.003	~ 0.004*
4-nitroaniline	1.00	0.000	~ 0.007
6-aminopurine	4.15	0.011	~ 0.009
aniline	4.606	0.012	~ 0.016
pyridine	5.22	0.014	~ 0.011
2-aminoquinoline	7.29	0.021	~ 0.025
4-aminopyridine	9.114	0.028	0.027
piperazine	9.79	0.030	~ 0.021
ethylamine	10.65	0.033	0.032
dimethylamine	10.776	0.033	0.029
pyrrolidine	11.305	0.035	0.032

*The approximate values (denoted $\sim$) may be in error by ± 0.005 (from Perrin, 1964).

two groups, is represented by $BH^+ \rightleftharpoons B + H^+$, and is described by the equations given above. The entropy term for the weaker group is assumed to be zero, which yields the relation

$$\frac{-d(pK_a)}{dT} = \frac{pK_a}{T} \qquad (11.11)$$

(T is temperature in degrees absolute). Some values calculated by this equation are given in Table 11.7.

Table 11.7   Temperature coefficients for divalent organic cations.

All pK values refer to the process

$$BH_2^{2+\prime} \rightleftharpoons BH^+ + H^+$$

Base	pK_a (25°C)	$-d(pK_a)/dT$	
		Calc.	Found
1, 2-diaminocyclohexane	6.34	0.021	~ 0.027*
ethylenediamine	6.85	0.023	0.027
1, 3-diamino-2-hydroxypropane	7.81	0.026	~ 0.025
1, 3-diaminopropane	8.49	0.028	~ 0.031
1, 4-diaminobutane	9.20	0.031	~ 0.030
1, 6-diaminohexane	9.83	0.033	0.034

*As in Table 11.6 (from Perrin, 1964).

The effect of temperature on the ionic product of water is given in Table 11.8. The effect of temperature on the pH of solutions of two commonly used standards (potassium hydrogen phthalate and sodium borate) is given in Table 11.9.

Table 11.8 Thermodynamic values of the ionic product of water (K_w)

°C	$K_w \times 10^{14}$	pK_w
0	0·1117	14·952
5	0·1820	14·740
10	0·2877	14·541
15	0·4446	14·352
20	0·6714	14·173
25	0·9908	14·004
30	1·439	13·842
35	2·042	13·690
40	2·851	13·545
45	3·917	13·407
50	5·297	13·276
55	7·079	13·150
60	9·311	13·031

(from Covington, Ferra and Robinson, 1977)

Table 11.9 The effect of temperature on pH of the standardizing buffers NBS Standard pH scale (Bates, 1962)

Temperature (°C)	0·05M potassium hydrogen phthalate pH	0·01M sodium borate (borax) pH
0	4·003	9·464
10	3·998	9·332
15	3·999	9·276
20	4·002	9·225
25	4·008	9·180
30	4·015	9·139
35	4·024	9·102
40	4·035	9·068
50	4·060	9·011
60	4·091	8·962

The third decimal figure is not significant.

Table 11.10   Calculation of percentage ionized

$pK_a - pH$	if Anion	if Cation	$pK_a - pH$	if Anion	if Cation
− 6·0	99·999 90	0·000 099 9	+ 0·1	44·27	55·73
− 5·0	99·999 00	0·000 999 9	+ 0·2	38·68	61·32
− 4·0	99·990 0	0·009 999 0	+ 0·3	33·39	66·61
− 3·5	99·968	0·031 6	+ 0·4	28·47	71·53
− 3·4	99·960	0·039 8	+ 0·5	24·03	75·97
− 3·3	99·950	0·050 1	+ 0·6	20·07	79·93
− 3·2	99·937	0·063 0	+ 0·7	16·63	83·37
− 3·1	99·921	0·079 4	+ 0·8	13·70	86·30
− 3·0	99·90	0·099 9	+ 0·9	11·19	88·81
− 2·9	99·87	0·125 7	+ 1·0	9·09	90·91
− 2·8	99·84	0·158 2	+ 1·1	7·36	92·64
− 2·7	99·80	0·199 1	+ 1·2	5·93	94·07
− 2·6	99·75	0·250 5	+ 1·3	4·77	95·23
− 2·5	99·68	0·315 2	+ 1·4	3·83	96·17
− 2·4	99·60	0·396 6	+ 1·5	3·07	96·93
− 2·3	99·50	0·498 7	+ 1·6	2·450	97·55
− 2·2	99·37	0·627 0	+ 1·7	1·956	98·04
− 2·1	99·21	0·787 9	+ 1·8	1·560	98·44
− 2·0	99·01	0·990	+ 1·9	1·243	98·76
− 1·9	98·76	1·243	+ 2·0	0·990	99·01
− 1·8	98·44	1·560	+ 2·1	0·787 9	99·21
− 1·7	98·04	1·956	+ 2·2	0·627 0	99·37
− 1·6	97·55	2·450	+ 2·3	0·498 7	99·50
− 1·5	96·93	3·07	+ 2·4	0·396 6	99·60
− 1·4	96·17	3·83	+ 2·5	0·315 2	99·68
− 1·3	95·23	4·77	+ 2·6	0·250 5	99·75
− 1·2	94·07	5·93	+ 2·7	0·199 1	99·80
− 1·1	92·64	7·36	+ 2·8	0·158 2	99·84
− 1·0	90·91	9·09	+ 2·9	0·125 7	99·87
− 0·9	88·81	11·19	+ 3·0	0·099 91	99·90
− 0·8	86·30	13·70	+ 3·1	0·079 4	99·921
− 0·7	83·37	16·63	+ 3·2	0·063 0	99·937
− 0·6	79·93	20·07	+ 3·3	0·050 1	99·950
− 0·5	75·97	24·03	+ 3·4	0·039 8	99·960
− 0·4	71·53	28·47	+ 3·5	0·031 6	99·968
− 0·3	66·61	33·39	+ 4·0	0·009 999 0	99·990 0
− 0·2	61·32	38·68	+ 5·0	0·000 999 9	99·999 00
− 0·1	55·73	44·27	+ 6·0	0·000 099 9	99·999 90
0	50·00	50·00			

APPENDIX V

How percentage ionized may be calculated, given pK_a and pH

The relevant equation for acids is (11.12), and for bases (11.13).

$$\% \text{Ionized} = \frac{100}{1 + 10^{(pK_a - pH)}} \qquad (11.12)$$

$$\% \text{Ionized} = \frac{100}{1 + 10^{(pH - pK_a)}} \qquad (11.13)$$

Results, obtained from these formulae, are given in Table 11.10.

APPENDIX VI

An outline of the theory of pH

The quantity known as 'pH' is defined as:

$$pH = - \log a_H \qquad (11.14)$$

in which a_H represents the activity of the hydrogen ion. pH is usually determined in a cell of the type represented by:

$$\text{Hg; Hg}_2\text{Cl}_2 | 3 \cdot 8\text{M KCl} \| \text{ test solution} | \text{glass electrode.} \qquad (11.15)$$

This is the accepted notation for a cell consisting of a calomel reference half cell, represented by 'Hg; Hg$_2$Cl$_2$|3·8M KCl' and a glass electrode dipping into the solution of which the pH is to be measured. The double vertical line represents the junction between (i) the reference half cell of invariant potential and (ii) the test solution, which allows electrical connection to be made between the two half cells via the bridge solution which, in this case, is 3·8M potassium chloride. This connection is often made by means of a porous plug sealed through the bottom of the vessel containing the electrode of calomel paste (which consists of mercury ground with mercury(I) chloride) and the above potassium chloride solution. As a result of the slow, unequal rates of diffusion of potassium ion and chloride ion through the porous plug into the test solution, a potential develops across the liquid junction (3·8M KCl and the test solution) the magnitude of which cannot be thermodynamically defined. Nonetheless, acknowledgement that this liquid junction potential exists is made by including, in the thermodynamically derived Nernst equation (11.16) for the cell, a term E_j to represent this indeterminate potential. Thus at 298 K (25°C):

$$E_{cell} = (E_G^0 - 0 \cdot 0592 \log a_{H^+}) - E_{cal} + E_j \qquad (11.16)$$

where E_G^0 is the standard electrode potential of the glass electrode over the time taken for the complete measurement, and E_{cal} is the potential of the calomel reference cell. In the measurement of pH (or indeed in any other ion activity measurement made with a cell of this type containing a liquid junction) there is no alternative but to assume that E_j will be constant for all measurements, and to express the equation (11.16) in a pseudo-thermodynamic form as:

$$E_{cell} = constant + slope \cdot pH. \qquad (11.17)$$

The term *constant* is thus a composite of E_G^0 and E_{cal} which have thermodynamic validity, and E_j which has not. For most correctly operating glass electrodes, the value *slope* is usually very close to that predicted by the value of the Nernst factor at the temperature of the determination (e.g. 0·0592 at 25°C), and most modern pH meters are equipped with a control which can be manipulated to ensure that the *slope* is, actually and numerically, equal to this Nernst factor during the course of the measurement.

The solving of equation (11.17) in terms of an unknown pH value, pH(X), requires that at least one calibration standard of known pH value, pH(S), be available. If the value of E_{cell} for this buffer solution of pH(S) is measured as E_S, and the electrodes are then transferred to the solution of unknown pH(X) for which the corresponding value of the potential E_X is measured, then the equations based on (11.17) can be written as:

$$E_S = constant + slope \cdot pH(S) \qquad \text{(for the buffer solution)}$$

and
$$E_X = constant + slope \cdot pH(X) \qquad \text{(for the test solution)}$$

from which:

$$pH(X) = pH(S) + (E_X - E_S)/slope. \qquad (11.18)$$

The assumption inherent in this equation is, of course, that the value of the liquid junction potential remains the same in both solutions, and that the value of *slope* remains constant over the time taken for the measurements.

The assignation of the pH(S) value to the buffer solution, however, presents difficulties because there is no way of accurately measuring, in a thermodynamically valid fashion, the activity of a single ionic species like the hydrogen ion, even with a perfectly functioning hydrogen electrode. Extra-thermodynamic assumptions must always be made as to the magnitude of the junction potential in the case of a cell with liquid junction, or to the magnitude of a single ion activity coefficient if a cell without liquid junction is used. The value of this type of activity coefficient cannot be determined experimentally.

In the formulation of the pH scale adopted by the National Bureau of Standards, Washington D.C., it was found preferable to use a cell without liquid junction for the measurement of cell potentials and to make an extra-thermodynamic assumption concerning the magnitude of a single ion activity coefficient. The type of cell used for this purpose can be represented as:

$$\text{Pt; } H_2(1 \text{ atm})|H^+(\text{pH buffer solution}), Cl^-(\text{known molality})|AgCl; Ag. \quad (11.19)$$

For the sake of brevity, we shall confine our discussion to a temperature of 298 K (25°C). At this temperature, the electrodes used in cell (11.19) can be calibrated by measuring E_{cell} for a 0·01m solution of hydrochloric acid for which the mean activity coefficient $\gamma_{\pm}$ (molal scale) has been determined as 0·9047. This allows the standard potential of the cell (E^0_{cell}), consisting of a hydrogen electrode and a silver–silver chloride electrode, to be determined because, for the cell reaction:

$$2AgCl_{(S)} + H_2 \,(1\text{ atm}) \rightleftharpoons 2Ag_{(S)} + 2H^+ + 2Cl^-,$$

the Nernst equation is:

$$E_{cell} = E^0_{cell} - \frac{0·059\,16}{2} \cdot \log a_H^2 \cdot a_{Cl}^2. \tag{11.20}$$

This equation can be written in terms of the activity of the hydrochloric acid electrolyte a_2 as:

$$E_{cell} = E^0_{cell} - 0·059\,16 \log a_2.$$

For a 1:1 electrolyte such as hydrochloric acid, a_2 is defined as:

$$a_2 = m_{HCl}^2 \gamma_{\pm}$$

and thus:

$$E_{cell} = E^0_{cell} - 0·118\,32 \log m_{HCl} \cdot \gamma_{\pm}^2.$$

For the known molality of hydrochloric acid (0·01m) and its mean activity coefficient (0·9047), E^0_{cell} can be evaluated from the measured potential of the cell by the equation:

$$E^0_{cell} = E_{cell} - 0·241\,79.$$

For solutions contained in cells of the type shown by cell (11.19), this calibration factor E^0_{cell} can be combined with the measured e.m.f. of the cell in order to define an acidity function for the solution. Thus from equation (11.20):

$$E_{cell} = E^0_{cell} - \frac{0·059\,16}{2} \cdot \log a_H^2 \cdot m_{Cl}^2 \cdot \gamma_{Cl}^2$$

which, for the known molality of chloride contained in the cell, can be written as:

$$(E_{cell} - E^0_{cell})/0·059\,16 + \log m_{Cl} = - \log a_H \gamma_{Cl}. \tag{11.21}$$

All the terms on the left-hand side of this equation are known; the thermodynamically valid term on the right-hand side is the acidity function, usually written as $p(a_H \gamma_{Cl})$.

This acidity function can be regarded as the thermodynamic precursor to the quasi-thermodynamic quantity known as pH. The pH of a given buffer solution is derived by measuring values of $p(a_H \gamma_{Cl})$ for a series of buffer solutions each of

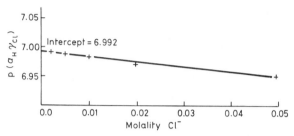

Figure 11.1 Plot of the acidity function against increasing chloride ion concentrations, for solutions containing 0·02 m KH_2PO_4 and 0·02 m Na_2HPO_4.

which contains the same concentration of buffer component(s) and different, but known, concentrations of chloride. A plot of the experimentally derived values of $p(a_H\gamma_{Cl})$ against the corresponding chloride concentration often gives a straight line relation which is extrapolated to $m_{Cl} = 0$ to yield a quantity designated as $p(a_H\gamma_{Cl})^0$. A typical plot obtained in our laboratories for solutions containing 0·02m potassium dihydrogen phosphate and 0·02m disodium hydrogen phosphate at the indicated chloride ion concentrations is given in Fig. 11.1. The intercept corresponds to the value of $p(a_H\gamma_{Cl})^0$ for this 0·02m phosphate buffer solution at 25°C: its pH is derivable from this value by equation (11.22) which employs the following expansion of the acidity function:

$$p(a_H\gamma_{Cl})^0 = -\log a_H - \log \gamma_{Cl}^0$$
$$pH = p(a_H\gamma_{Cl})^0 = \log \gamma_{Cl}^0. \qquad (11.22)$$

In order to calculate pH by this equation, an extra-thermodynamic assumption about the magnitude of the activity coefficient for the chloride ion must now be made, and for this purpose the Debye–Hückel equation (see equation (3·4)) is used in the form appropriate for concentrations expressed in mol kg^{-1} as:

$$-\log \gamma_{Cl}^0 = 0·511\, I^{1/2}/(1 + 1·5I^{1/2}). \qquad (11.23)$$

In this Bates–Guggenheim version of the equation pertinent to 25°C, the value of the ion size parameter is assumed to be 0·45 nm; the uncertainty introduced by this value, even at low ionic strengths, causes the uncertainty in the value of γ_{Cl}^0 and thus in the derived pH of the buffer solution. For the example of Fig. 11.1, the ionic strength of the 0·02m phosphate solution at zero chloride ion concentration is 0·08m, and hence the pH value for this buffer solution, given by combining equations (11.22) and (11.23), is

$$pH = 6·992 - 0·511\,(0·08)^{1/2}/[1 + 1·5\,(0·08)^{1/2}]$$
$$= 6·891.$$

The pH values for most of the buffer solutions given in Table 2.1 were assigned using this method, and the uncertainty of the values is estimated to be $\pm 0·006$ pH at 25°C and somewhat larger at higher temperatures.

In summary, therefore, it is important to realize that although pH can be measured quite precisely (± 0.002), its accuracy is possibly not greater than ± 0.01 unit for solutions similar in composition to the buffer solutions. In solutions whose ionic strengths are low compared to those of these buffer solutions, the magnitude of the liquid junction potential is likely to increase, and the accuracy in the measurement of pH is expected to decrease, as the general nature of the test solution becomes increasingly divergent from that of the standardizing buffer solution. The contribution of the liquid junction potential is also likely to be increased when measuring the pH of solutions of pH < 3 and pH > 11, due to the comparatively high mobilities of the hydrogen and hydroxyl ions, respectively.

References

Adams, E.Q. (1916) *J. Amer. Chem. Soc.*, **38**, 1503.

Adams, R. and Mahan, J.E. (1942) *J. Amer. Chem. Soc.*, **64**, 2588.

Albert, A. (1953) *Biochem. J.*, **54**, 646.

Albert, A. (1955) *J. Chem. Soc.*, 2690.

Albert, A. (1963) *Physical Methods in Heterocyclic Chemistry*, Vol. 1 (ed. Katritzky, A.R.), Academic Press, New York, p. 2.

Albert, A. (1966a) *The Acridines*, 2nd edn, Edward Arnold, London.

Albert, A. (1966b) *J. Chem. Soc. B*, 427.

Albert, A. (1967) *Angewand. Chem., Internat. Edn.*, **6**, 919.

Albert, A. (1971) *Physical Methods in Heterocyclic Chemistry*, Vol. 3 (ed. A.R. Katritzky), Academic Press, New York, p. 1.

Albert, A. (1973) in *Synthetic Procedures in Nucleic Acid Chemistry* (eds. W.W. Zorbach and R.S. Tipson, Wiley-Interscience, New York.

Albert, A. (1976) *Adv. Hetrocycl. Chem.* **20**, 117.

Albert, A. (1979) *Selective Toxicity*, 6th edn, Chapman and Hall, London; Methuen, New York.

Albert, A. and Armarego, W.L.F. (1965) *Adv. Heterocycl. Chem.* **4**, 1.

Albert, A. and Serjeant, E.P. (1960) *Biochem. J.*, **76**, 621.

Albert, A., Goldacre, R.J. and Phillips, J.N. (1948) *J. Chem. Soc.*, 2240.

Ang, K.P. (1959) *J. Chem. Soc.*, 3822.

Angyal, S.J. and Angyal, C.L. (1952) *J. Chem. Soc.* 1461.

Arnett, E.M. and Bushick, R.D. (1965) *J. Amer. Chem. Soc.*, **86**, 1564.

Arnett, E.M. and Mach, G.W. (1964) *J. Amer. Chem. Soc.*, **86**, 2671.

Arnett, E.M. and Wu, C.Y. (1960) *J. Amer. Chem. Soc.*, **82**, 4999.

Arnett, E.M. and Wu, C.Y. (1962) *J. Amer. Chem. Soc.*, **84**, 1680.

Arnett, E.M., Quirk, R. and Burke, J. (1970) *J. Amer. Chem. Soc.*, **92**, 1260.

Arnett, E.M., Quirk, R. and Larsen, J. (1970) *J. Amer. Chem. Soc.*, **92**, 3977.

Avdeef, A. and Bucher, J.J. (1978) *Anal. Chem.*, **50**, 2137.

Avedikian, L. (1966) *Bull. Soc. chim., France*, 2570.

Ballinger, P. and Long, F.A. (1959) *J. Amer. Chem. Soc.*, **81**, 1050.

Ballinger, P. and Long, F.A. (1960) *J. Amer. Chem. Soc.*, **82**, 795.

Barker, J.A. and Beecham, A.F. (1960) *Austral. J. Chem.*, **13**, 1.

Barlin, G.B. and Perrin, D.D. (1972) in *Technique of Organic Chemistry* Vol. 4, Part 1 (ed. A. Weissberger) Wiley-Interscience, New York.

Bartgel, J. (1976) *Thermometric Titrations*, Wiley-Interscience, New York.

Bartlett, P.D., Roha, M. and Stiles, R.M. (1954) *J. Amer. Chem. Soc.* **76**, 2349.

Bascombe, K.N. and Bell, R.P. (1959) *J. Chem. Soc.*, 1096.

Bates, R.G. (1962) *J. Res. Natl. Bureau Standards*, **66A**, 179.

Bates, R.G. (1973) *Determination of pH, Theory and Practice*, 2nd edn, Wiley, New York.

Bates, R.G. and Gary, R. (1961) *J. Res. Natl. Bureau Standards*, **65A**, 495.

Bates, R.G., Pinching, G.D. and Smith, E.R. (1950) *J. Res. Natl. Bureau Standards*, **45**, 418.

Beke, D. (1963) *Adv. Heterocycl. Chem.*, **1**, 167.

Bell, R.P. (1960) *The Proton in Chemistry*, Methuen, London.

Bell, R.P. (1969) *Acids and Bases*, 2nd edn, Methuen, London.

Bellingham, P. Johnson, C.D. and Katritzky, A.R. (1967) *J. Chem. Soc. B*, 1226.

Belotserkovskaya, N.G. and Ginsberg, O.F. (1964) *Doklady Akad. Nauk SSSR*, **155**, 1098.

Benesch, R.E., and Benesch, R. (1955) *J. Amer. Chem. Soc.*, **77**, 5877.

Benet, L.Z. and Goyan, J.E. (1967) *J. Pharm. Sci.*, **56**, 665.

Bissot, T.C., Parry, R.W., and Campbell, D.H. (1957) *J. Amer. Chem. Soc.*, **79**, 796.

Bjerrum, J. (1941). *Metal Ammine Formation in Aqueous Solution*, Haase, Copenhagen.

Bjerrum, N. (1923) *Zeits. physik. Chem.*, **104**, 147.

Bolton, P.D., Fleming, K.A. and Hall, F.M. (1972) *J. Amer. Chem. Soc.*, **94**, 1033.

Bolton, P.D. and Hall, F.M. (1969) *J. Chem. Soc. B*, 1047.

Bolton, P.D., Hall, F.M. and Reece, I.H. (1966) *Spectrochim. Acta*, **22**, 1149.

Bonner, O.D. and Torres A.L. (1965) *J. Phys. Chem.*, **69**, 4109.

Bonner, T.G. and Phillips, J. (1966) *J. Chem. Soc. B*, 650.

Bonvicini, P., Levi, A., Lucchini, V., Modena, G. and Scorrano, G. (1973) *J. Amer. Chem. Soc.*, **95**, 5960.

Bordwell, F.G. and Cooper, G.D. (1952) *J. Amer. Chem. Soc.*, **74**, 1058, 6317.

Bowden, K. (1966) *Chem. Rev.*, **66**, 119.

Brignell, P.J., Johnson, C.D., Katritzky, A.R., Shakir, N., Tarkan, H.O. and Walker, G. (1967) *J. Chem. Soc. B*, 1233.

Brockman, F.G. and Kilpatrick, M. (1934) *J. Amer. Chem. Soc.*, **56**, 1483.

Brønsted, J.N. (1923) *Receuil Trav. chim. Pays-Bas*, **42**, 718.

Brown, D.J. and Ghosh, P. (1969) *J. Chem. Soc. B*, 270.

Brown, D.J., Hoerger, E., and Mason, S.F. (1955) *J. Chem. Soc.*, 4035.

Brown, H.C., Bartholomay, H., and Taylor, M.D. (1944) *J. Amer. Chem. Soc.*, **66**, 435.

Bryson, A. (1951) *Trans. Farad. Soc.*, **47**, 522.

Bryson, A., Davies, N.R. and Serjeant, E.P. (1963) *J. Amer. Chem. Soc.*, **85**, 1933.

Bryson, A. and Matthews, R.W. (1961) *Austral. J. Chem.*, **14**, 237.

Bunting, J.W. and Meathrel, W.G. (1972) *Canad. J. Chem.*, **50**, 917.

Bunting, J.W. (1979) *Adv. Heterocycl. Chem.*, **25**, 1.

Campbell, H.J. and Edward. J.T. (1960) *Canad. J. Chem.*, **38**, 2109.

Catlin, W.E. (1935) *Iowa State Coll. J. Sci.* **10**, 65 (C.A. **1936**, 30, 935).

Cavill, G.W.K., Gibson, N.A., and Nyholm, R.S. (1949) *J. Chem. Soc.*, 2466.

Chiang, Y. and Whipple, E.B. (1963) *J. Amer. Chem. Soc.*, **85**, 2763.

Christensen, J.J., Izatt, R.M., Hansen, L.D. and Partridge, J.A. (1966) *J. Phys. Chem.*, **70**, 2003.

Christensen, J.J., Rytting, J.H. and Izatt, R.M. (1970) *J. Chem. Soc. B*, 1646.

Clark, J. and Cunliffe, A.E. (1973) *Chem. Ind.*, 281.

Clarke, J.H.R. and Woodward, L.A. (1966) *Trans. Farad. Soc.*, **62**, 2226.

Conant, J.B. and Wheland, G.W. (1932) *J. Amer. Chem. Soc.*, **54**, 1212.

Cookson, R.F. (1974) *Chem. Rev.*, **74**, 5.

Cordes, E.H. and Jencks, W.P. (1962) *J. Amer. Chem. Soc.*, **84**, 833.

Covington, A.K., Ferra, M.I.A. and Robinson, R.A. (1977) *J. Chem. Soc. Farad. Trans.* *1* **73**, 1721.
Covington, A.K. and Lilley, T.H. (1967) *Trans. Farad. Soc.*, **63**, 1749.
Debye, P. and Hückel, E. (1923) *Physikal. Zeit.*, **24**, 185, 305, 334.
Debye, P. and Hückel, E. (1924) *Physikal. Zeit.*, **25**, 145.
Deno, N.C. and Wisotsky, M.J. (1963) *J. Amer. Chem. Soc.*, **85**, 1735.
Dessy, R.E., Kitching, W., Psarras, T., Salinger, R., Chen, A. and Chivers, T. (1966) *J. Amer. Chem. Soc.*, **88**, 460.
Dickson, T.R. (1968) *The Computer and Chemistry, an Introduction to Programming and Numerical Methods*, Freeman, San Francisco, p. 141.
Dinius, R.H. and Choppin, G.R. (1962) *J. Phys. Chem.*, **66**, 268.
Dippy, J.F.J. (1939) *Chem. Rev.*, **25**, 151.
Doty, P. and Ehrlich, E. (1952) *Ann. Rev. Phys. Chem.*, **3**, 81.
Ebel, H.T. (1969) *Die Acidität der CH-Saüren*, Georg Thieme Verlag, Stuttgart (A reprint of a chapter from Houben-Weyl's *Methoden der Organischen Chemie*, Vol. XIII-I).
Ebert, L. (1926) *Z. physik. Chem.*, **121**, 385.
Edsall, J.T. and Blanchard, M.H. (1933) *J. Amer. Chem. Soc.*, **55**, 2337.
Edsall, J.T., Martin, R.B. and Hollingworth, B.R. (1958) *Proc. Natl. Acad. Sci., U.S.*, **44**, 505.
Edward, J.T., Chang, H.S., Yates, K. and Stewart, P. (1960) *Canad. J. Chem.*, **38**, 1518.
Edward, J.T., Leane, J.B. and Wang, I.C. (1962) *Canad. J. Chem.*, **40**, 1521.
Farmer, R.C. and Warth, F.J. (1904) *J. Chem. Soc.*, **85**, 1713.
Fraenkel, G. and Franconi, C. (1960) *J. Amer. Chem. Soc.*, **82**, 4478.
Fuoss, R.M. and Kraus, C.A. (1933) *J. Amer. Chem. Soc.*, **55**, 476.
Goldacre, R.J. and Phillips, J.N. (1949) *J. Chem. Soc.*, 1724.
van der Graaf, B., Hoefnagel, A.J. and Wepster, B.M. (1981) *J. Org. Chem.*, **46**, 653.
Green, R.W. and Tong, H.K. (1956) *J. Amer. Chem. Soc.*, **78**, 4896.
Greig, C.C. and Johnson, C.D. (1968) *J. Amer. Chem. Soc.*, **90**, 6453.
Grimison, A., Ridd., J.H. and Smith, B.V. (1960) *J. Chem. Soc.*, 1352.
Grunwald, E. and Berkowitz, B.J. (1951) *J. Amer. Chem. Soc.*, **73**, 4939.
Gutbezahl, B. and Grunwald, E. (1953) *J. Amer. Chem. Soc.*, **75**, 559.
v. Halban, H. and Brüll, J. (1944) *Helv. Chim. Acta*, **27**, 1719.
Hall, N.F. (1930) *J. Amer. Chem. Soc.*, **52**, 5115.
Hall, N.F. and Sprinkle, M.R. (1932) *J. Amer. Chem. Soc.*, **54**, 3469.
Hammett, L.P. (1970) *Physical Organic Chemistry*, 2nd edn, McGraw-Hill, New York.
Hammett, L.P. and Deyrup, A.J. (1932) *J. Amer. Chem. Soc.*, **54**, 2721.
Handloser, C.S., Chakrabarty, M.B. and Mosher, M.W. (1973) *J. Chem. Educ.*, **50**, 510.
Hansch, C. (1971) in *Drug Design*, Vol. I (ed. E. Ariens), Academic Press, New York, p. 271.
Harned, H.S. and Ehlers, R.W. (1932) *J. Amer. Chem. Soc.*, **54**, 1350.
Harned, H.S. and Hecker, J.C. (1933) *J. Amer. Chem. Soc.*, **55**, 4838.
Harned, H.S. and Owen, B.B. (1958) *Physical Chemistry of Electrolytic Solutions*, 3rd edn, Reinhold, New York.
Harned, H.S. and Robinson, R.A. (1940) *Trans. Farad. Soc.*, **36**, 973.
Hartley, F.R., Burgess, C. and Alcock, R. (1980) *Solution Equilibria*, Ellis Horwood, Chichester, UK.
Hietanen, S. and Sillén, L.G. (1952) *Acta Chem. Scand.*, **6**, 747.
Hills, G.J. and Ives, D.J.G. (1951) *J. Chem. Soc.*, 305.
Hildebrand, J.H. (1913) *J. Amer. Chem. Soc.*, **35**, 847.

Hinman, R.L. and Lang, J. (1964) *J. Amer. Chem. Soc.*, **86**, 3796.

Högfeldt, E. (1982) *Stability Constants of Metal-Ion Complexes, Part A-Inorganic Ligands*, Compiled for IUPAC, Pergamon, Oxford.

Högfeldt, E. and Bigeleisen, J. (1960) *J. Amer. Chem. Soc.*, **82**, 15.

Hood, G.C. and Reilly, C.A. (1960) *J. Chem. Phys.*, **32**, 127.

International Union of Pure and Applied Chemistry (1978) *Pure Appl. Chem.*, **50**, 1485.

Irving, H.M. and Bell, C.F. (1952) *J. Chem. Soc.*, 1216.

Irving, H.M. and Rossotti, H.S. (1953) *J. Chem. Soc.*, 3397.

Irving, H.M., Rossotti, H.S. and Harris, C. (1955) *Analyst*, **80**, 83.

Irving, H.M. and Williams, R.J.P. (1950) *J. Chem. Soc.*, 2890.

Ives, D.J.G. and Pryor, J.H. (1955) *J. Chem. Soc.*, 2104.

Jameson, R.F. and Wilson, M.F. (1972) *J. Chem. Soc. Dalton Trans.*, 2607.

Jaques, D. and Leisten, J. (1964) *J. Chem. Soc.*, 2683.

Jones, G. and Prendergast, M.J. (1937) *J. Amer. Chem. Soc.*, **59**, 731.

Jones, J.R. (1973) *The Ionization of Carbon Acids*, Academic Press, London.

Johnson, C.D., Katritzky, A.R., Ridgewell, B.J., Shakir, N. and White, A.M. (1965) *Tetrahedron*, **21**, 1055.

Jukes, T.H. and Schmidt, C.L.A. (1935) *J. Biol. Chem.*, **110**, 9.

Katritzky, A.R. and Waring, A.J. (1962) *J. Chem. Soc.*, 1540.

Katritzky, A.R., Waring, A.J. and Yates, K. (1963) *Tetrahedron*, **19**, 465.

King, E.J. (1965) *Acid-Base Equilibria*, Pergamon, Oxford.

King, E.J. and Prue, J.E. (1961) *J. Chem. Soc.*, 275.

Klabunde, K.J. and Burton, D.J. (1972) *J. Amer. Chem. Soc.*, **94**, 820.

Koeberg-Telder, A. and Cerfontain, H. (1975) *J. Chem. Soc. Perkin II*, 226.

Kolthoff, I.M. (1925) *Biochem. Zeit.*, **162**, 289.

Kolthoff, I.M. (1927) *J. Amer. Chem. Soc.*, **49**, 1218.

Kolthoff, I.M. and Laitinen, H.A. (1941) *pH and Electro-titrations*, Wiley, New York.

Kolthoff, I.M., Lingane, J.J. and Larson, W.D. (1938) *J. Amer. Chem. Soc.*, **60**, 2512.

Kortüm, G., Vogel, W. and Andrussow, K. (1961) *Dissociation Constants of Organic Acids in Aqueous Solution* (compiled for IUPAC), Butterworths, London.

Krebs, H.A. and Speakman, J.C. (1945) *J. Chem. Soc.*, 593.

Landini, D., Modena, G., Scorrano, G. and Taddei, I. (1969) *J. Amer. Chem. Soc.*, **91**, 6703.

Lee, D.G. (1970) *Canad. J. Chem.*, **48**, 1919.

Lee, D.G. and Cameron, R. (1971) *J. Amer. Chem. Soc.*, **93**, 4724.

Leggett, D.J. (ed.) (1983) *Computational Methods for the Determination of Stability Constants*, Plenum, New York.

Levy, G.C., Cargioli, J.D. and Racela, W. (1970) *J. Amer. Chem. Soc.*, **92**, 6238.

de Ligny, C.L., Loriaux, H. and Ruiter, A. (1961) *Rec. Trav. chim. Pays-Bas*, **80**, 725.

Lister, M.W. (1955) *Canad. J. Chem.*, **33**, 426.

Lumme, P.O. (1957) *Suomen Kemistil. (Finland)*, **30**, B, 173.

Lundén, H. (1907) *J. Chim. physique*, **5**, 574.

MacInnes, D.A. and Shedlovsky, T. (1932) *J. Amer. Chem. Soc.*, **54**, 1429.

MacInnes, D.A., Shedlovsky, T. and Longsworth, L.G. (1932) *J. Amer. Chem. Soc.*, **54**, 2758.

Markham, R. and Smith, J.D. (1951) *Nature, Lond.*, **168**, 406.

Martin, R.B. (1971) *J. Phys. Chem.*, **75**, 2657.

Martin, R.B. and Edsall, J.T. (1958) *Bull. Soc. Chim. biol., France*, **40**, 1763.

Martindale, W. (1982) *The Extra Pharmacopoeia*, 28th edn (ed. J. Reynolds) Pharmaceutical Press, London.

Mellander, A. (1934) *Svensk. kem. Tidskr.*, **46**, 99.

Milward, A.F. (1969) *Analyst, Lond.*, **94**, 154.

Mizutani, M. (1925) *Zeits. physik. Chem.*, **118**, 318, 327.

Mukherjee, L.M. and Grunwald, E. (1958) *J. Phys. Chem.*, **62**, 1311.

Murray, M.A. and Williams, G. (1950) *J. Chem. Soc.*, 3322.

Noyce, D.S. and Jorgenson, M.J. (1962) *J. Amer. Chem. Soc.*, **84**, 4312.

Ogston, A.G. and Peters, R.A. (1936) *Biochem. J.*, **30**, 736.

Onsager, L. (1926) *Physikal. Zeit.*, **27**, 388.

Onsager, L. (1927) *Physikal. Zeit.*, **28**, 277.

Ostwald, W. (1889) *Zeit. physik. Chem.*, **3**, 170.

Pascual, C. and Simon, W. (1964) *Helv. Chim. Acta*, **47**, 683.

Paul, M.A. and Long, F.A. (1957) *Chem. Rev.*, **57**, 1.

Pecsok, R.L., Shields, L.D., Cairns, T. and McWilliam, J. (1976) *Modern Methods of Chemical Analysis*, 2nd edn, Wiley, New York.

Pedersen, K.J. (1943) *Kgl. danske videnskab. Selskab, Maths. fys. Medd.*, **20**, 7 (C.A. 1944, **38**, 4854).

Perrin, D.D. (1960) *J. Chem. Soc.*, 3189.

Perrin, D.D. (1963) *Austral. J. Chem.*, **16**, 572.

Perrin, D.D. (1964) *Austral. J. Chem.*, **17**, 484.

Perrin, D.D. (1965a) *Dissociation Constants of Organic Bases in Aqueous Solution* (compiled for IUPAC), Butterworths, London.

Perrin, D.D. (1965b) *Adv. Heterocycl. Chen.*, **4**, 43.

Perrin, D.D. (1972) *Dissociation Constants of Organic Bases in Aqueous Solution*, First Supplement (Compiled for IUPAC), Pergamon, Oxford.

Perrin, D.D. (1979) *Stability Constants of Metal-Ion Complexes, Part B-Organic Ligands*, Second Supplement (compiled for IUPAC), Pergamon, Oxford.

Perrin, D.D. (1983) *Dissociation Constants of Inorganic Acids and Bases*, 2nd edn, (Compiled for IUPAC), Butterworths, London.

Perrin, D.D. and Dempsey, B. (1974) *Buffers for pH and Metal Ion Control*, Chapman and Hall, London; Methuen, New York.

Perrin, D.D., Dempsey, B. and Serjeant, E.P. (1981) pK_a *Prediction for Organic Acids and Bases*, Chapman and Hall, London; Methuen, New York.

Perrin, D.D. and Sayce, I.G. (1966) *Chem. Ind.*, 661.

Pfeiffer, P. (1907) *Ber. deutsch. chem. Ges.*, **40**, 4036.

Pleskov, V.A. and Monoszon, A.M. (1935) *Zhur. fiz. Khim.*, **6**, 513.

Pressman, D. and Brown, D.H. (1943) *J. Amer. Chem. Soc.*, **65**, 540.

Pring, J.N. (1924) *Trans. Farad. Soc.*, **19**, 705.

Rabenstein, D.L. and Anvarhusein, A.I. (1982) *Anal. Chem.*, **54**, 526.

Reagen, M.T. (1969) *J. Amer. Chem. Soc.*, **91**, 5506.

Reeves, R.L. (1966) *J. Amer. Chem. Soc.*, **88**, 2240.

Ricketts, J.A. and Cho, C.S. (1961) *J. Org. Chem.*, **26**, 2125.

Robinson, R.A. and Biggs, A.I. (1957) *Austral. J. Chem.*, **10**, 128.

Robinson, R.A. and Bower, V.E. (1961) *J. Phys. Chem.*, **65**, 1279.

Robinson, R.A. and Stokes, R.H. (1959) *Electrolyte Solutions*, 2nd edn, Butterworths, London.

Rochester, C.H. (1966) *Quart. Rev. Chem. Soc., London*, **20**, 511.

Rochester, C.H. (1970) *Acidity Functions*, Academic Press, London.

Rossotti, F.J.C. and Rossotti, H.S. (1961) *The Determination of Stability Constants*, McGraw Hill, New York, p. 146.

Ryabova, R.S., Medvetskaya, I.M. and Vinnik, M.I. (1966) *Russ. J. Phys. Chem.*, **40**, 182.

Samén, E. (1947) *Arkiv. Kemi Mineral. Geol.*, **24** B, No. 6 (C.A. 1948, **42**, 6313).

Sayce, I.G. (1968) *Talanta*, **15**, 1397.

Schwarzenbach, G. and Lutz, K. (1940) *Helv. Chim. Acta.*, **23**, 1147, 1162.

Schwarzenbach, G. and Sulzberger, R. (1944) *Helv. Chim. Acta*, **27**, 348.

Searles, S.S., Tamres, M.T., Block, F. and Quarterman, L.A. (1956) *J. Amer. Chem. Soc.*, **78**, 4917.

Serjeant, E.P. (1964) personal observation.

Serjeant, E.P. (1969) *Austral. J. Chem.*, **22**, 1189.

Serjeant, E.P. (1984) *Potentiometry and Potentiometric Titrations*, Wiley, New York.

Serjeant, E.P. and Dempsey, B. (1979) *Ionization Constants of Organic Acids in Aqueous Solution* (compiled for IUPAC), Pergamon, Oxford (a supplement to book listed under Kortüm, G. *et al.*).

Serjeant, E.P. and Warner, A.G. (1978) *Anal. Chem.*, **50**, 1724.

Shedlovsky, T. (1932) *J. Amer. Chem. Soc.*, **54**, 1411.

Shedlovsky, T. (1938) *J. Franklin Institute*, **225**, 739.

Shedlovsky, T. (1962) in *Electrolytes*, (ed. B. Pesce) Pergamon, New York, p. 146.

Shedlovsky, T. and Shedlovsky, L. (1971) in *Techniques of Chemistry*, Vol. 1: *Physical Methods of Chemistry*, Part IIA, (eds. A. Weissberger and B.W. Rossiter) Wiley, New York, p. 180.

Sillén, L.G. (1964) *Acta Chem. Scand.*, **18**, 1085.

Sillén, L.G. and Martell, A.E. (1964) *Stability Constants of Metal-Ion Complexes*, Compiled for IUPAC, Special Publication No. 17, The Chemical Society, London.

Sillén, L.G. and Martell, A.E. (1971) *Stability Constants of Metal-Ion Complexes*, Supplement No. 1, Special Publication No. 25, The Chemical Society, London.

Simon, W. (1964) *Angew. Chem., Internat. Edn. Engl.*, **2**, 661.

Sinistri, C. and Villa, L. (1962) *Il Farmaco, Ed. sci., Italy*, **17**, 949.

Skvortsov, N.K., Dogadino, A.V., Tereshchenko, G.T., Morkovin, N.V., Ionin, B.I. and Petrov, A.H. (1971) *Zhur. obshchei Khim* **41**, 2807.

Speakman, J.C. (1940) *J. Chem. Soc.*, 855.

Spillane, W., Hannigan, T. and Shelly, K. (1982) *J. Chem. Soc. Perkin II*, 19.

Spillane, W. and Thomson, J.B. (1977) *J. Chem. Soc. Perkin II*, 580.

Stearn, A.E. (1931) *J. Phys. Chem.*, **35**, 2226.

Stewart, R. and O'Donnell, J.P. (1962) *J. Amer. Chem. Soc.* **84**, 493.

Stroh, H.-H. and Westphal, G. (1963) *Chem. Ber.*, **96**, 184.

Swain, C. and Lupton, E. (1968) *J. Amer. Chem. Soc.*, **90**, 4328.

Taft, R.W. (1960) *J. Phys. Chem.*, **64**, 1803, 1805.

Taft, R.W. and Levine, P.L. (1962) *Anal. Chem.*, **34**, 436.

Tate, M.E. (1981) *Biochem. J.*, **195**, 419.

Thamsen, J. (1952) *Acta Chem. Scand.*, **6**, 270.

Tickle, P., Briggs, A.G. and Wilson, J.M. (1970) *J. Chem. Soc. B*, 65.

Virtanen, P.O.I. and Korpela, J. (1968) *Suomen Kemistil., (Finland)*, B, **41**, 321, 326 (C.A. 1969, **70**, 41 339).

Voerman, G.L. (1907) *Rec. Trav. chim. Pays-Bas*, **26**, 293.

Wegscheider, R. (1895) *Monatsh.*, **16**, 153.

Wegscheider, R. (1902) *Monatsh.*, **23**, 287.
Wepster, B.M. (1952) *Rec. Trav. chim. Pays-Bas*, **71**, 1159, 1171.
Werner, A. (1907) *Ber. deutsch. chem. Ges.*, **40**, 272.
Weston, R.E., Ehrenson, S. and Heizinger, K. (1967) *J. Amer. Chem. Soc.*, **89**, 481.
Williams, G. and Hardy, M.L. (1953) *J. Chem. Soc.*, 2560.
de Wit (1982) *Ann. Biochem.*, **123**, 285.
Yates, K., Stevens, J.B. and Katritzky, A.R. (1964) *Canad. J. Chem.*, **42**, 1957.
Young, T.F., Wu, Y.C. and Krawetz, A.A. (1957) *Discuss. Farad. Soc.*, **24**, 37.

Index